T0182079

EAI/Springer Innovations in Communication and Computing

Series editor
Imrich Chlamtac, CreateNet, Trento, Italy

Editor's Note

The impact of information technologies is creating a new world yet not fully understood. The extent and speed of economic, life style and social changes already perceived in everyday life is hard to estimate without understanding the technological driving forces behind it. This series presents contributed volumes featuring the latest research and development in the various information engineering technologies that play a key role in this process.

The range of topics, focusing primarily on communications and computing engineering include, but are not limited to, wireless networks; mobile communication; design and learning; gaming; interaction; e-health and pervasive healthcare; energy management; smart grids; internet of things; cognitive radio networks; computation; cloud computing; ubiquitous connectivity, and in mode general smart living, smart cities, Internet of Things and more. The series publishes a combination of expanded papers selected from hosted and sponsored European Alliance for Innovation (EAI) conferences that present cutting edge, global research as well as provide new perspectives on traditional related engineering fields. This content, complemented with open calls for contribution of book titles and individual chapters, together maintain Springer's and EAI's high standards of academic excellence. The audience for the books consists of researchers, industry professionals, advanced level students as well as practitioners in related fields of activity include information and communication specialists, security experts, economists, urban planners, doctors, and in general representatives in all those walks of life affected ad contributing to the information revolution.

About EAI

EAI is a grassroots member organization initiated through cooperation between businesses, public, private and government organizations to address the global challenges of Europe's future competitiveness and link the European Research community with its counterparts around the globe. EAI reaches out to hundreds of thousands of individual subscribers on all continents and collaborates with an institutional member base including Fortune 500 companies, government organizations, and educational institutions, provide a free research and innovation platform.

Through its open free membership model EAI promotes a new research and innovation culture based on collaboration, connectivity and recognition of excellence by community.

More information about this series at http://www.springer.com/series/15427

Ju Bin Song • Husheng Li • Marceau Coupechoux
Editors

Game Theory for Networking Applications

 Springer

Editors
Ju Bin Song
Department of Electronic Engineering
Kyung Hee University
Yongin, Korea (Republic of)

Husheng Li
Electrical Engineering/Computer Sciences
The University of Tennessee
Knoxville, TN, USA

Marceau Coupechoux
Telecom ParisTech
Paris, France

ISSN 2522-8595   ISSN 2522-8609   (electronic)
EAI/Springer Innovations in Communication and Computing
ISBN 978-3-030-06575-1   ISBN 978-3-319-93058-9   (eBook)
https://doi.org/10.1007/978-3-319-93058-9

This Springer imprint is published by the registered company Springer Nature Switzerland AG
The registered company address is: Gewerbestrasse 11, 6330 Cham, Switzerland

Preface

Game theory has recently become a useful tool for modeling and studying various networks. The past decade has witnessed a huge explosion of interest in issues that intersect networks and game theory. With the rapid growth of data traffic, from any kind of devices and networks, game theory is requiring more intelligent transformation. Game theory is called to play a key role in the design of new generation networks that are distributed, self-organizing, cooperative, and intelligent.

This book consists of invited and technical papers of GAMENETS 2018, and contributed chapters on game theoretic applications such as networks, social networks, and smart grid.

Yongin, Republic of Korea　　　　　　　　　　　　　　　　Ju Bin Song
Knoxville, TN, USA　　　　　　　　　　　　　　　　　　　Husheng Li
Paris, France　　　　　　　　　　　　　　　　　Marceau Coupechoux
April 2018

Contents

Contributors

S. M. Ahsan Kazmi Department of Computer Science and Engineering, Kyung Hee University, Yongin, South Korea

Eitan Altman Univ Côte d'Azur, INRIA, Sophia Antipolis, Méditerranée, France

LINCS, Paris, France

Chen Avin Ben-Gurion University of the Negev, Beer-Sheba, Israel

Kaigui Bian School of Electronics Engineering and Computer Science, Peking University, Beijing, China

Mariia Bulgakova Saint Petersburg State University, Saint Petersburg, Russia

Ramya Burra Department of ESE, Indian Institute of Science, Bangalore, India

Avi Cohen Weizmann Institute of Science, Rehovot, Israel

Elena Gubar St. Petersburg State University, Faculty of Applied Mathematics and Control Processes, Saint-Petersburg, Russia

Zhu Han Department of Electrical and Computer Engineering, University of Houston, Houston, TX, USA

Department of Computer Science and Engineering, Kyung Hee University, Seoul, South Korea

Choong Seon Hong Department of Computer Science and Engineering, Kyung Hee University, Yongin, South Korea

Changhee Joo Ulsan National Institute of Science and Technology (UNIST), Ulsan, South Korea

Sungwook Kim Department of Computer Science, Sogang University, Seoul, South Korea

Song-Kyoo Kim Abu Dhabi School of Management, Abu Dhabi, United Arab Emirates

Alexei Korolev National Research University Higher School of Economics, St. Petersburg, Russia

Alexander Yu. Krylatov Saint Petersburg State University, Saint-Petersburg, Russia

Solomenko Institute of Transport Problems, St. Petersburg, Russia

Joy Kuri Department of ESE, Indian Institute of Science, Bangalore, India

Elena A. Lezhnina Saint Petersburg State University, Saint-Petersburg, Russia

Zvi Lotker Ben-Gurion University of the Negev, Beer-Sheba, Israel

Peter Marbach University of Toronto, Toronto, ON, Canada

Vladimir Matveenko National Research University Higher School of Economics, St. Petersburg, Russia

Hung Khanh Nguyen University of Houston, Houston, TX, USA

David Peleg Weizmann Institute of Science, Rehovot, Israel

Leon Petrosyan Saint Petersburg State University, Saint Petersburg, Russia

Ivan V. Popov Control Methods and Robotics, TU Darmstadt, Darmstadt, Germany

Neetu Raveendran Department of Electrical and Computer Engineering, University of Houston, Houston, TX, USA

Abylay Satybaldy Ulsan National Institute of Science and Technology (UNIST), Ulsan, South Korea

Artem Sedakov Saint Petersburg State University, Saint Petersburg, Russia

Chandramani Singh Department of ESE, Indian Institute of Science, Bangalore, India

Ju Bin Song Department of Electronic Engineering, Kyung Hee University, Yongin, South Korea

Lingyang Song School of Electronics Engineering and Computer Science, Peking University, Beijing, China

Vladislav Taynitskiy St. Petersburg State University, Faculty of Applied Mathematics and Control Processes, Saint-Petersburg, Russia

Nguyen H. Tran Department of Computer Science and Engineering, Kyung Hee University, Yongin, South Korea

Haitao Xu University of Science and Technology Beijing, Beijing, China

Victor V. Zakharov Saint Petersburg State University, Saint-Petersburg, Russia

Qingchao Zeng School of Economics and Management, Beihang University, Beijing, People's Republic of China

Larry Yueli Zhang University of Toronto, Toronto, ON, Canada

Xianwei Zhou University of Science and Technology Beijing, Beijing, China

Quanyan Zhu Department of Electrical and Computer Engineering, Tandon School of Engineering, New York University, Brooklyn, NY, USA

Acronyms

ACK	ACKnowledgement
AODV	Ad hoc On-demand Distance Vector
CC	Cloud Computing
CDIM	Credit Distribution and Influence Maximization
CP	Content Provider
CR	Cognitive Radio
CSI	Channel State Information
CSSC	Cloud and SNS Supported Collaboration
CU	Cellular User
DCI	Downlink Control Information
DoS	Denial-of-Service
DSRP	Distributed Secure Routing Protocol
EU	End User
EV	Electric Vehicle
FTP	File Transfer Protocol
GS	Gale-Shapley
GS	General Sharing
HARQ	Hybrid Automatic Repeat reQuest
HetNet	Heterogeneous Network
HM	Hierarchical Matching Algorithm
ICT	Information and Communications Technology
IDP	Imputation Distribution Procedure
ILP	Integer Linear Programming
IMUG	Influence Maximization for Unknown Graphs
InP	Infrastructure Provider
IoT	Internet of Things
ISP	Internet Service Provider
KKT	Karush–Kuhn–Tucker
LOP	Link Optimized Payoff
LTE	Long-Term Evolution
MBS	Macro Base Station

MR	Maximal Recursive
MUE	Macrocell User Equipment
MVNO	Mobile Virtual Network Operator
NACK	NACKnowledgement
NECD	Noncooperative Energy Charging and Discharging
NOMA	Non-Orthogonal Multiple Access
NP	Nonlinear Programming
NS	Network Simulator
OFDMA	Orthogonal Frequency Division Multiple Access
OMA	Orthogonal Multiple Access
PBCH	Physical Broadcast CHannel
PDCCH	Physical Downlink Control CHannel
PEV	Plug-in Electric Vehicle
PER	Packet Error Rate
PHEV	Plug-in Hybrid Electric Vehicle
PHICH	Physical HARQ Indicator CHannel
PPP	Poisson Point Process
PU	Primary User
PUCCH	Physical Uplink Control CHannel
PUE	Primary User Emulation
QoS	Quality-of-Service
RA	Resource Allocation
RB	Resource Block
SBS	Small Cell Station
SC	Social Cloud
SED	Square Euclidean Distance
SIC	Successive Interference Cancellation
SINR	Signal to Interference plus Noise Ratio
SIR	Susceptible-Infected-Recovered
SN	Social Network
SNS	Social Network Service
SR	Scheduling Request
SU	Secondary User
SUE	Small cell User Equipment
TCP	Transmission Control Protocol
TTI	Transmission Time Interval
UB	Upper Bound
UCI	Uplink Control Information
UE	User Equipment
V2B	Vehicle-to-Building
V2G	Vehicle-to-Grid
V2X	Vehicle-to-everything
WNV	Wireless Network Virtualization

Part I
Game Theory for Networks

Chapter 1
Types of Nodes and Centrality Measures in Networks

Vladimir Matveenko and Alexei Korolev

1.1 Introduction

In network games, players' behavior in equilibrium is defined by their positions in the network, which are described by one or other centrality measure. However, different centrality measures prove important in different models: in particular, degree centrality—in a model of systematic shift of social norms [7], betweenness centrality—in a Medici of Florence power case (see [6]), eigenvalue centrality—in a model of aggregation of information by society [5], Katz-Bonacich centrality—in a model of criminal behavior [1], diffusion centrality—in models of diffusion in networks [2], a special case of alpha centrality (hereinafter referred to as alpha-gamma centrality)—in a model of production of knowledge [10, 11]. Such variety of measures of centrality found to be of importance in various situations leads naturally to open challenging questions about the nature of the centrality measures themselves. There is an emergent literature on the interrelations of different centrality measures and their relation with other structural characteristics of networks (e.g., [3, 8, 9]).

In the present paper we show that there is a set of centrality measures which characterize not just separate nodes of networks, but also types of nodes selected by a rather universal structural characteristic—a network typology. The concept of network typology is based on the fact that in any network (undirected graph) $\langle M, N \rangle$ with a set of nodes $M = \{1, \ldots, n\}$ and a set of edges M, the nodes may be colored in a minimal number of colors, S, in such a way that any node of color $i = 1, \ldots, S$ has a definite number, t_{ij}, of neighbors of color j, $i = 1, \ldots S$. Such coloring provides a division of the set of nodes M into S types. Correspondingly,

V. Matveenko (✉) · A. Korolev
The National Research University Higher School of Economics, St. Petersburg, Russia
e-mail: vmatveenko@hse.ru

© Springer Nature Switzerland AG 2019
J. B. Song et al. (eds.), *Game Theory for Networking Applications*,
EAI/Springer Innovations in Communication and Computing,
https://doi.org/10.1007/978-3-319-93058-9_1

to each network a "type adjacency matrix" corresponds: $T = (t_{ij})$ of size $S \times S$. Two networks are said to have the same typology if they have the same type adjacency matrix T.

The concept of network typology differs of other approaches to the structure of networks. We will see, for instance, that networks of similar size and similar typology may have different topological structure. In the same time, networks of similar size and similar distribution of degrees may have different typologies.

We consider a class of several centrality measures (degrees, eigenvector centrality, Katz-Bonacich centrality, diffusion centrality, alpha-gamma centrality, and alpha-beta centrality). The majority of them are familiar to the reader (see [6]). Alpha-gamma centrality and alpha-beta centrality are special cases of alpha centrality. The definitions follow. The vector of α centralities of nodes is

$$C^\alpha = (I - \alpha A)^{-1} h, \qquad (1.1)$$

where I is the identity matrix, A is the adjacency matrix, α is a number (may be negative), h is a vector. Bonacich [4] introduced a special case of the alpha centrality, called alpha-beta centrality[1]:

$$C^{\alpha\beta} = \beta(I - \alpha A)^{-1} A 1, \qquad (1.2)$$

where 1 is the vector of all ones, α and β are parameters; may be positive or negative. Another special case of alpha centrality—alpha-gamma centrality—was found by Matveenko and Korolev [10, 11] in a model of knowledge production in a network. The vector of alpha-gamma centralities of nodes is

$$C^{\alpha\gamma} = \gamma(I - \alpha A)^{-1} 1, \qquad (1.3)$$

where α and γ are parameters such that $\alpha\gamma < 0$.

We show that the centrality measures of our class are defined by the network typology, i.e. depend on types of nodes. This implies that networks of different size but with the same typology have common properties in many games. In particular, this implies that game equilibria corresponding to any of these centrality measures may be transplanted among networks of the same typology.

We show that information, needed for calculation of the above-mentioned centrality measures for nodes of various types, relates only on knowledge of typology, but knowledge of the full structure of the network is not needed. Thus, the information containing in the type adjacency matrix T is enough for calculating the listed centrality measures. Notice that the size of the type adjacency matrix T may be several orders lower than the size of the adjacency matrix A.

---

[1]Bonacich [4] uses the letters α, β in the opposite order. We rewrite the definition to obtain a formula of the same type as familiar definition of the Katz-Bonacich centrality.

Opposite to some other centrality measures, the Katz-Bonacich centrality, the alpha-gamma centrality and the alpha-beta centrality exist only in some regions of parameter α. The Katz-Bonacich centrality exists under $\alpha \in (0, 1/\lambda_F)$, where λ_F is the Frobenius eigenvalue of both the adjacency matrix A and the type adjacency matrix T. For the alpha-gamma centrality and the alpha-beta centrality the existence conditions are more complex. Thus, possibilities to subjectively choose parameter α are rather limited. For example, for the Katz-Bonacich centrality to be defined for star network, parameter α has to decline unlimitedly when the star network grows. The Bonacich centrality and the alpha-gamma centrality never coexist under a joint α.

Konig et al. [9] and Bloch et al. [3] formulate the following problem: for which classes of networks each centrality measure of a set of centrality measures defines the same order on the set of nodes of network? In fact, two questions are here: about the class of networks and about the set of centrality measures. In particular, Bloch et al. [3] introduces a class of trees—so-called regular monotonous hierarchies—and proves that, for any tree of this class, orders defined on the set of nodes by the following set of centrality measures: degree centrality, decay centrality, Katz-Bonacich centrality, diffusion centrality, intermediary centrality—do coincide. For each tree not belonging the class, there exists a pair of centralities from listed above which defines different orders on the set of nodes.

We show that such classes of networks are not limited by trees. We prove that for any typology with two types of nodes, the orders defined on the set of nodes by the following centrality measures: degree, eigenvalue centrality, Katz-Bonacich centrality, diffusion centrality, alpha-gamma centrality, and alpha-beta centrality—do coincide. The intersection of the classes of networks described in [3] and in our paper is the class of star networks.

The rest of the paper is organized in the following way. In Sect. 1.2 we introduce the concepts of type of node and network typology. In Sect. 1.3 we establish a relation between the network typology and several centrality measures. In Sect. 1.4 we consider classes of networks, for which orders generated by several different centrality measures do coincide. Section 1.5 concludes.

1.2 Types of Nodes and Network Topology

We consider a network (undirected graph) $\langle M, N \rangle$, where $M = \{1, \ldots, n\}$ is a set of nodes and N is a set of edges. Our concept of network typology is based on the fact that the nodes may be colored in a minimal number of colors, S, in such a way that any node of color $j (j = 1, \ldots, S)$ has definite numbers of neighbors of each of S colors. Such coloring provides a division of the set of nodes N into S types.

More formally, a set of types of the nodes of the network is the minimal set $I = \{1, \ldots, S\}$ for which there exists a mapping $f : M \to I$ such that if $i, j \in M$

and $f(i) = f(j)$, then $f(v(i)) = f(v(j))$, where $v(i)$ is a set of neighbors of node i. A polynomial algorithm of division of the set of nodes into types is described in [10, 11].

To each network a "type adjacency matrix" corresponds. It is a matrix $T = (t_{ij})$ of size $S \times S$, where t_{ij} is the number of neighbors of type j for any node of type i. Two networks are said to have the same typology if they have the same type adjacency matrix.

A class of one-type networks is the familiar class of regular networks (i.e., such that all nodes have the same degree). The next in the order of complexity is the class of nodes with 2 types of nodes. It is a subclass of the class of networks with two degrees.

Each tree possesses a unique typology. For each network which is not a tree there is a sequence of networks of the same typology with increasing number of nodes. Figure 1.1 demonstrates three networks of different size with the same typology.

Figure 1.2 shows that networks of similar size (6 nodes) and similar typology ($T = \begin{pmatrix} 1 & 2 \\ 1 & 1 \end{pmatrix}$) may have different topological structure: in particular, the network in Fig. 1.2a has a bridge (an edge, after removing of which the number of connected components increases), while there are no bridges in the network in Fig. 1.2b. The average clusterings of these networks are also different: 2/3 for the network in Fig. 1.2a and 0 for the network in Fig. 1.2b.

It is also easy to show that networks of a similar size and a similar distribution of degrees may have different typologies. An important structural property of networks is provided by the following Lemma.

Lemma 1 *Let i, j be types of nodes, which have, correspondingly, n_i and n_j nodes. If $t_{ij} \neq 0$, then $t_{ji} \neq 0$ and*

$$n_i t_{ij} = n_j t_{ji}, \qquad (1.4)$$

Proof Each of products in (1.4) expresses the number of all edges connecting nodes of type i with nodes of type j.   □

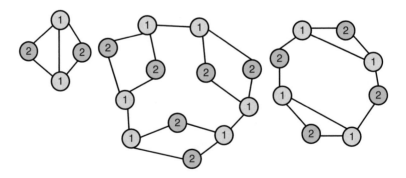

Fig. 1.1 Networks of different size with the same typology

Fig. 1.2 Networks of the
same typology and size but
with different topologies

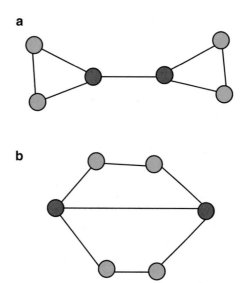

Networks belonging to the same typology possess some common properties defined by the type adjacency matrix. Let n_i be the number of type i nodes in a network $(i = 1, \ldots, S)$, m_d be the number of nodes of degree d, $(d \in D)$ where D is the set of different degrees of nodes in the network.

Theorem 1 *The following statistics are invariants of typology (they are the same for any networks of given typology):*

1. *relative numbers of nodes of different types:* n_i/n_j $(i \neq j;\; i, j = 1, \ldots, S)$;
2. *distribution of nodes by types:* n_i/n $(i = 1, \ldots, S)$;
3. *distribution of nodes by degrees:* m_d/n $(d \in D)$;
4. *relative degree:* $\frac{\sum_{d \in D} m_d d}{n}$.

Proof It is implied by Lemma 1. □

1.3 Relation of Typology and Centralities

Theorem 2 *In any class of networks with same typology, if i and j are nodes of the same type (may be even belonging different networks of this typology), then $c(i) = c(j)$, where c is any of the following centrality measures: degree, eigenvalue centrality, Katz-Bonacich centrality, diffusion centrality, alpha-gamma centrality, alpha-beta centrality.*

This implies that networks of different size but with the same typology have common properties; in particular, game equilibria related to the centrality measures may be transplanted among networks with the same typology.

Theorem 2 is equivalent to the following Theorem 3.

Theorem 3 *For calculation of any of the centrality measures listed in Theorem 2, the type adjacency matrix T may be used instead of the adjacency matrix A.*

Proof For degree, Katz-Bonacich centrality, diffusion centrality, alpha-gamma centrality, alpha-beta centrality, it rather easily follows from corresponding definitions. Let us prove it for eigenvalue centrality. It follows directly from the definitions that (*i*) if λ is an eigenvalue of the type adjacency matrix T and $\tilde{b}$ is a corresponding eigenvector, then λ is also an eigenvalue of the adjacency matrix A, and the corresponding eigenvector is b, where $b_i = \tilde{b}_{\tilde{\imath}}$ if $i \in M_{\tilde{\imath}}$. (In other words, $\lambda\tilde{b} = T\tilde{b}$ implies $\lambda b = Ab$). Let us prove that (ii) if λ_F is the Frobenius eigenvalue of the type adjacency matrix T, then λ_F is also the Frobenius eigenvalue of the adjacency matrix A.

To prove (ii) ad absurdum, assume that the Frobenius eigenvalue of matrix A is $\mu \neq \lambda_F$. Part (i) implies that $\mu > \lambda_F$. Let e be the Frobenius eigenvector of matrix A and let $\hat{e}$ be the n-vector with components

$$\hat{e}_i = \max_{j \in M_i} e_j, \ i = 1, 2, \ldots, n.$$

Evidently,

$$A\hat{e} \geq \mu e.$$

Let $\hat{f}$ be size S vector corresponding to $\hat{e}$, i.e. $\hat{f}_{\tilde{\imath}} = \hat{e}_i$ if $i \in M_{\tilde{\imath}}$. Then

$$T\hat{f} \geq \mu\hat{f}, \tag{1.5}$$

but according to the Perron-Frobenius theorem, since λ_F is the Frobenius eigenvalue, (1.5) implies $\lambda_F \geq \mu$ Contradiction!   □

1.4 Classes of Networks, for Which Orders Generated by Several Different Centrality Measures Do Coincide

Konig et al. [9] and Bloch et al. [3] formulate the following problem: for which classes of networks each of a set of several centrality measures defines the same order on the set of nodes of network? (Here are two questions: about the class of networks and about the set of centrality measures). In particular, Bloch et al. [3] introduces a class of trees—so called regular monotonous hierarchies. Applying to undirected networks, the class of regular monotonous hierarchies might be defined in the following way. Let $\rho(i)$ be a distance between a node i and a root of the tree, and d_i be degree of node i.

Definition Tree g is called *regular monotonous hierarchy* if there exists a node i_0 (root) such that the tree satisfies the following conditions:

- All nodes being at the same distance from the root have the same degree (i.e., if $\rho(i) = \rho(j)$, then $d_i = d_j$).
- For any two nodes i, j, if the distance between the root and i, $\rho(i)$, is less than the distance between the root and j, $\rho(j)$, then $d_i \geq d_j$.

Bloch et al. [3] proves that if a tree g is a regular monotonous hierarchy, then the following centrality measures: degree centrality, decay centrality, Katz-Bonacich centrality, diffusion centrality, intermediary centrality order the nodes in the same way. For each tree not of the class of regular monotonous hierarchies, there is a pair of centralities from this list, which will give different orders on the set of nodes.

We show that such kind classes of networks are not limited by trees.

Theorem 4 *For any typology with two types of nodes, the orders on the set of nodes, defined by the following centrality measures: degree, eigenvalue centrality, Katz-Bonacich centrality (under condition of existence), diffusion centrality, alpha-gamma centrality (under condition of existence and a positive determinant of matrix $(I - \alpha T)$, $\Delta > 0$), and alpha-beta centrality (under conditions $\beta \Delta > 0, d_1 > \alpha \Delta, d_2 > \alpha \Delta$) do coincide.*

Proof (Degrees and Eigenvector Centrality) The vector $\tilde{C}^e$ of eigenvector centralities is defined by the equation

$$T\tilde{C}^e = \alpha_F \tilde{C}^e,$$

which implies

$$t_{11}\tilde{C}_1^e + t_{12}\tilde{C}_2^e = \frac{t_{11} + t_{22} + \sqrt{(t_{11} - t_{22})^2 + 4t_{12}t_{21}}}{2}\tilde{C}_1^e.$$

Hence,

$$t_{12}\tilde{C}_2^e = \left(\frac{t_{11} + t_{22} + \sqrt{(t_{11} - t_{22})^2 + 4t_{12}t_{21}}}{2} - t_{11}\right)\tilde{C}_1^e,$$

$$\tilde{C}_1^e > \tilde{C}_2^e \Leftrightarrow 2t_{12} + t_{11} - t_{22} > \sqrt{(t_{11} - t_{22})^2 + 4t_{12}t_{21}}$$

$$\Leftrightarrow 4t_{12}^2 + 4t_{12}(t_{11} - t_{22}) > 4t_{12}t_{21} \Leftrightarrow d_1 > d_2.$$

$\square$

Proof (Degrees and Bonacich Centrality) Let us proof that

$$\tilde{C}_1^B - \tilde{C}_2^B = \frac{1}{\alpha\left(\frac{1}{\alpha} - \lambda_f\right)\left(\frac{1}{\alpha} - \lambda_2\right)}(d_1 - d_2). \tag{1.6}$$

Indeed,

$$\tilde{C}_1^B - \tilde{C}_2^B = \frac{1}{\Delta}[1 + \alpha(t_{12} - t_{22}) - 1 - \alpha(t_{21} - t_{11})] = \frac{\alpha}{\Delta}(d_1 - d_2),$$

$$\Delta = 1 - \alpha T_r T + \alpha^2 Det\, T = \alpha^2 \left(\frac{1}{\alpha^2} - \frac{1}{\alpha} T_r T + Det\, T \right)$$

$$= \alpha^2 \left(\frac{1}{\alpha} - \lambda_F \right) \left(\frac{1}{\alpha} - \lambda_2 \right).$$

The Bonacich centrality exists iff

$$\frac{1}{\alpha} > \lambda_F \geq \lambda_2,$$

hence, as it is seen from (1.6), the orders do coincide.   □

Proof (Degrees and Diffusion Centrality) We will use the definition of diffusion centrality with a free term:

$$C^{dif} = (I + \alpha T + \alpha^2 T^2 + \cdots + \alpha^L T^L)\tilde{1}. \tag{1.7}$$

If $L = 1$, then the values of diffusion centrality of types are

$$C_i^{dif} = 1 + \alpha d_{\tilde{i}}.$$

The value α is positive; thus, if $L = 1$, then the diffusion centralities induce on the set of nodes the same order as degrees of nodes. Let us show that in a network with two types of nodes, the order of nodes induced by diffusion centrality does not depend on the natural number L. For the type adjacency matrix,

$$T = \begin{pmatrix} t_{11} & t_{12} \\ t_{21} & t_{22} \end{pmatrix}, \tag{1.8}$$

the eigenvalues are

$$\lambda_F = \frac{t_{11} + t_{22} + \sqrt{(t_{11} - t_{22})^2 + 4t_{12}t_{21}}}{2},$$

$$\lambda_2 = \frac{t_{11} + t_{22} - \sqrt{(t_{11} - t_{22})^2 + 4t_{12}t_{21}}}{2}.$$

Let

$$\Lambda = \begin{pmatrix} \lambda_F & 0 \\ 0 & \lambda_2 \end{pmatrix}, \tag{1.9}$$

and let C be a matrix of transition to a basis constructed from the eigenvectors of matrix T; i.e. columns of matrix C are eigenvectors corresponding eigenvalues λ_F and λ_2. Elements of matrix C will be denoted c_{ij}, and elements of the inverse matrix C^{-1} will be denoted $\tilde{c}_{ij}$. Then:

$$(I + \alpha T + \alpha^2 T^2 + \cdots + \alpha^L T^L)\tilde{1}$$

$$= C(I + \alpha \Lambda + \alpha_2 \Lambda^2 + \cdots + \alpha_L \Lambda^L)C^{-1}\tilde{1}$$

$$= \begin{pmatrix} c_{11} & c_{12} \\ c_{21} & c_{22} \end{pmatrix} * \begin{pmatrix} \frac{1-\lambda_F^{L+1}}{1-\lambda_F} & 0 \\ 0 & \frac{1-\lambda_2^{L+1}}{1-\lambda_2} \end{pmatrix} * \begin{pmatrix} \tilde{c}_{11} & \tilde{c}_{12} \\ \tilde{c}_{21} & \tilde{c}_{22} \end{pmatrix} * \begin{pmatrix} 1 \\ 1 \end{pmatrix}$$

$$= \begin{pmatrix} c_{11}\frac{1-\lambda_F^{L+1}}{1-\lambda_F} & c_{12}\frac{1-\lambda_2^{L+1}}{1-\lambda_2} \\ c_{21}\frac{1-\lambda_F^{L+1}}{1-\lambda_F} & c_{22}\frac{1-\lambda_2^{L+1}}{1-\lambda_2} \end{pmatrix} * \begin{pmatrix} \tilde{c}_{11} & \tilde{c}_{12} \\ \tilde{c}_{21} & \tilde{c}_{22} \end{pmatrix} * \begin{pmatrix} 1 \\ 1 \end{pmatrix}$$

$$= \begin{pmatrix} \tilde{c}_{11}\tilde{c}_{11}\frac{1-\lambda_F^{L+1}}{1-\lambda_F} + c_{12}\tilde{c}_{21}\frac{1-\lambda_2^{L+1}}{1-\lambda_2} & c_{11}\tilde{c}_{12}\frac{1-\lambda_F^{L+1}}{1-\lambda_F} + c_{12}\tilde{c}_{22}\frac{1-\lambda_2^{L+1}}{1-\lambda_2} \\ c_{21}\tilde{c}_{11}\frac{1-\lambda_F^{L+1}}{1-\lambda_F} + c_{22}\tilde{c}_{21}\frac{1-\lambda_2^{L+1}}{1-\lambda_2} & c_{21}\tilde{c}_{12}\frac{1-\lambda_F^{L+1}}{1-\lambda_F} + c_{22}\tilde{c}_{22}\frac{1-\lambda_2^{L+1}}{1-\lambda_2} \end{pmatrix}\begin{pmatrix} 1 \\ 1 \end{pmatrix}$$

$$= \begin{pmatrix} c_{11}\tilde{c}_{11}\frac{1-\lambda_F^{L+1}}{1-\lambda_F} + c_{12}\tilde{c}_{21}\frac{1-\lambda_2^{L+1}}{1-\lambda_2} + c_{11}\tilde{c}_{12}\frac{1-\lambda_F^{L+1}}{1-\lambda_F} + c_{12}\tilde{c}_{22}\frac{1-\lambda_2^{L+1}}{1-\lambda_2} \\ c_{21}\tilde{c}_{11}\frac{1-\lambda_F^{L+1}}{1-\lambda_F} + c_{22}\tilde{c}_{21}\frac{1-\lambda_2^{L+1}}{1-\lambda_2} + c_{21}\tilde{c}_{12}\frac{1-\lambda_F^{L+1}}{1-\lambda_F} + c_{22}\tilde{c}_{22}\frac{1-\lambda_2^{L+1}}{1-\lambda_2} \end{pmatrix}$$

Hence, the difference of the 1st and 2nd types nodes diffusion centralities is

$$Dif = \frac{1-\lambda_F^{L+1}}{1-\lambda_F}(\tilde{c}_{11} + \tilde{c}_{12})(c_{11} - c_{21}) + \frac{1-\lambda_2^{L+1}}{1-\lambda_2}(\tilde{c}_{21} + \tilde{c}_{22})(c_{12} - c_{22}).$$

This formula is true for any natural L, and also for $L = 0$. But for $L = 0$, obviously, $Dif = 0$; hence,

$$(\tilde{c}_{11} + \tilde{c}_{12})(c_{11} - c_{21}) + (\tilde{c}_{21} + \tilde{c}_{22})(c_{12} - c_{22}) = 0.$$

If degrees of the 1st type nodes are higher than of the 2nd type nodes, then for $L = 1$ we have $Dif > 0$, which implies

$$(\tilde{c}_{11} + \tilde{c}_{12})(c_{11} - c_{21}) > 0, (\tilde{c}_{21} + \tilde{c}_{22})(c_{12} - c_{22}) < 0.$$

But the difference of diffusion centralities has then the same sign for any other natural $L > 1$, as

$$\frac{1 - \lambda_F^{L+1}}{1 - \lambda_F} = 1 + \lambda_F + \lambda_F^2 + \cdots + \lambda_F^L,$$

$$\frac{1 - \lambda_2^{L+1}}{1 - \lambda_2} = 1 + \lambda_2 + \lambda_2^2 + \cdots + \lambda_2^L,$$

$$1 + \lambda_F + \lambda_F^2 + \cdots + \lambda_F^L > 0,$$

$$1 + \lambda_F + \lambda_F^2 + \cdots + \lambda_F^L > |1 + \lambda_2 + \lambda_2^2 + \cdots + \lambda_2^L|,$$

i.e. if degrees of the 1st type nodes are higher than of the 2nd type nodes, then $Dif > 0$ for any L. □

Proof (Degrees and Alpha-Gamma Centrality) The vector of $\alpha\gamma$ centralities of types is

$$\tilde{C}^{\alpha\gamma} = \frac{\alpha}{\Delta} \begin{pmatrix} 1 + \alpha(t_{12} - t_{22}) \\ 1 + \alpha(t_{21} - t_{11}) \end{pmatrix} = \begin{pmatrix} \frac{\gamma}{\Delta} + \frac{\alpha\gamma}{\Delta}(t_{12} - t_{22}) \\ \frac{\gamma}{\Delta} + \frac{\alpha\gamma}{\Delta}(t_{21} - t_{11}) \end{pmatrix}.$$

As $\alpha\gamma < 0$, if $\Delta < 0$, then

$$\tilde{C}_1^{\alpha\gamma} > \tilde{C}_2^{\alpha\gamma} \Leftrightarrow t_{12} - t_{22} > t_{21} - t_{11} \Leftrightarrow d_1 > d_2.$$

If $\Delta < 0$, then vice versa,

$$\tilde{C}_1^{\alpha\gamma} > \tilde{C}_2^{\alpha\gamma} \Leftrightarrow d_1 > d_2.$$

□

Proof (Degrees and Alpha-Beta Centrality) We have

$$\tilde{1} - \alpha T = \begin{pmatrix} 1 - \alpha t_{11} & -\alpha t_{12} \\ -\alpha t_{21} & 1 - \alpha t_{22} \end{pmatrix},$$

$$\Delta = 1 - \alpha t_{11} - \alpha t_{22} + \alpha^2 t_{11} t_{22} - \alpha^2 t_{12} t_{21},$$

$$\beta(\tilde{1} - \alpha T)^{-1} = \frac{\beta}{\Delta} \begin{pmatrix} 1 - \alpha t_{22} & -\alpha t_{12} \\ \alpha t_{21} & 1 - \alpha t_{11} \end{pmatrix},$$

$$\beta(\tilde{1} - \alpha T)^{-1} T = \frac{\beta}{\Delta} \begin{pmatrix} t_{11} - \alpha t_{11} t_{22} + \alpha t_{12} t_{21} & t_{12} \\ t_{21} & \alpha t_{12} t_{21} + t_{22} - \alpha t_{11} t_{22} \end{pmatrix},$$

$$\tilde{C}_{\alpha\beta} = \beta(\tilde{I} - \alpha T)^{-1}T1 = \frac{\beta}{\Delta}\begin{pmatrix} t_{11} + t_{12} - \alpha(t_{11}t_{22} - t_{12}t_{21}) \\ t_{21} + t_{22} - \alpha(t_{11}t_{22} - t_{12}t_{21}) \end{pmatrix}.$$

Thus, the $\alpha\beta$ centrality exists if

$$\beta\Delta > 0; t_{11}t_{22} > \alpha(t_{11}t_{22} - t_{12}t_{21}),$$

$$t_{21} + t_{22} > \alpha(t_{11}t_{22} - t_{12}t_{21})$$

or

$$\beta\Delta < 0; t_{11}t_{22} < \alpha(t_{11}t_{22} - t_{12}t_{21}),$$

$$t_{21} + t_{22} < \alpha(t_{11}t_{22} - t_{12}t_{21}).$$

In the former case, the $\alpha\beta$ centrality defines on the set of nodes the same order as degrees, and in the latter case the opposite order. □

1.5 Conclusion

Games on networks provide a productive way to analyze social and economic networks. Numerous research studies show that behaviors of players in equilibrium depend on network structure and, in particular, on positions of players in network, described by centrality measures. However, the concepts of network structure and player's position do themselves need a deeper understanding. Commonly, a game equilibrium is described by one or another particular centrality measure, but it is not quite clear, what is the reason of correspondence of a particular centrality measure to a particular game.

In the present paper we introduce two fundamental structural concepts—types of nodes and network typology—and show their importance in analysis of game equilibria. We find that centrality measures which belong to a class (consisting of degree, eigenvalue centrality, Katz-Bonacich centrality, diffusion centrality, alpha-gamma centrality, and alpha-beta centrality) do correspond, in fact, not to separate nodes but to types of nodes. An important consequence is that if, in a game equilibrium, behaviors of players are described by one of these centrality measures, then the equilibrium may be transplanted to any network possessing the same typology, independently on the size of network or on topological characteristics, such as presence or absence of bridges. The transplantation means copying behavior by players of the same type. Thus, in many games, the equilibrium behavior is defined primarily by the network typology and types of nodes, while centrality measures as structural characteristics of games are found to be secondary relatively to the types of nodes.

Another, more specific, problem discussed in the paper is the existence of a class of networks for which all centrality measures of a class do generate the same order of nodes. Staying in such class of networks, it is easy to compare equilibria of different games. We find that any typology with two types of nodes has such property for our class of network centralities (degree, eigenvalue centrality, Katz-Bonacich centrality, diffusion centrality, alpha-gamma centrality, and alpha-beta centrality). For any fixed typology with two types of nodes, all the above-mentioned centrality measures provide the same order on the set of nodes.

Thus, we characterize a class of games for which it is proven by different authors that the equilibrium strategies of players are described by one or another of the above-mentioned centrality measures. For this class of games there is a correspondence between typologies with two types of nodes and game equilibria. In future work, it would be important to find any other joint properties characterizing explicitly these games as representatives of one joint class

Acknowledgements The research is supported by the Russian Foundation for Basic Research (project 17-06-00618).

References

1. Ballester, C., Calvo-Armengol, A., Zenou, Y.: Who's who in networks. Wanted: the key player. Econometrica **74**, 1403–1417 (2006)
2. Banerjee, A., Chandrasekhar, A., Duflo, E., Jackson, M.O.: Diffusion of microfinance. Science **341**, 1236498 (2013)
3. Bloch, F., Jackson, M.O., Tebaldi, P.: Centrality measures in networks (2017). ArXiv: 1608.05845v3
4. Bonacich, P.B.: Power and centrality: a family of measures. Am. J. Sociol. **92**, 1170–1182 (1987)
5. Golub, B., Jackson, M.O.: Naïve learning in social networks and the wisdom of crowds. Am. Econ. J. Macroecon. **2**(1), 112–149 (2010)
6. Jackson, M.O.: Social and Economic Networks. Princeton University Press, Princeton (2008)
7. Jackson, M.O.: The friendship paradox and systematic biases in perceptions and social norms (2017). ArXiv: 1605.04470
8. Jackson, M.O.: A typology of social capital and associated network measures (2018). https://arxiv.org/abs/1711.09504
9. Konig, M., Tessone, C., Zenou, Y.: Nestedness in networks: a theoretical model and some applications. Theor. Econ. **9**(3), 695–752 (2014)
10. Matveenko, V.D., Korolev, A.V.: Equilibria in networks with production and knowledge externalities. In: Kalyagin, V.A., Koldanov, P.A., Pardalos, P.M. (eds.) Models, Algorithms and Technologies for Network Analysis. Springer Proceedings in Mathematics and Statistics, vol. 156, pp. 291–331. Springer, Cham (2016)
11. Matveenko, V.D., Korolev, A.V.: Knowledge externalities and production in network: game equilibria, types of nodes, network formation. Int. J. Comput. Econ. Econ. **7**(4), 323–358 (2017)

Chapter 2
The Time-Consistent Shapley Value for Two-Stage Network Games with Pairwise Interactions

Leon Petrosyan, Mariia Bulgakova, and Artem Sedakov

2.1 Introduction

Network games is a new and important part of modern game theory. Networks illustrate the interaction of both individuals and groups. For the first time in the literature, a non-cooperative form of pairwise interaction in a network was considered in [3] meaning direct interactions between network neighbors. Finding an equilibrium in online gaming as an example of a Designer–Adversary game was described in [4]. Pairwise interaction was exposed in [1] on the example of the dissemination of information and misinformation in social networks. The efficiency and stability of networks depending on external factors such as marginal costs were examined in [6]. An approach for finding optimal behavior in multistage games was considered in [9]. Cooperation in network games and a model of interaction between coalitions were considered in [5].

When cooperative behavior is investigated, it is important that players follow a cooperative agreement during the whole game. If a solution of the cooperative game is time-consistent, players have no reason to deviate from the accepted agreement. An imputation distribution procedure (IDP) which is a payment scheme that provides the implementation of the solution was introduced in [8] to prevent players from deviating from the cooperative agreement. The conditions for the time consistency of the core for two-stage games with pairwise interactions were established in [2]. The dynamic properties of cooperative solutions in multicriteria games were considered in [7]. In this paper, we provide analytic expressions for characteristic functions in a two-stage game with pairwise interactions. Further, similar to [11], we provide conditions for the time consistency of the Shapley value

L. Petrosyan · M. Bulgakova (✉) · A. Sedakov
Saint Petersburg State University, Saint Petersburg, Russia
e-mail: l.petrosyan@spbu.ru; a.sedakov@spbu.ru

© Springer Nature Switzerland AG 2019
J. B. Song et al. (eds.), *Game Theory for Networking Applications*,
EAI/Springer Innovations in Communication and Computing,
https://doi.org/10.1007/978-3-319-93058-9_2

in this game. Moreover we simplify the formula of the Shapley value for a network of a special type—a star.

2.2 Description of the Model

Let N be a finite set of players who make decisions in two stages, $|N| = n \geq 2$. At the first stage z_1 each player $i \in N$ chooses his behavior $b_i^1 = (b_{i1}^1, \ldots, b_{in}^1)$—a profile of offers to establish connections with other players:

$$b_{ij}^1 = \begin{cases} 1, & \text{if } j \in M_i, \\ 0, & \text{otherwise}, \end{cases}$$

with

$$\sum_{j \in N} b_{ij}^1 \leq a_i.$$

Here $M_i \subseteq N \setminus \{i\}$ is a given set of players whom player i can offer connections, $b_{ii}^1 = 0$ for $i \in N$; $a_i \in \{0, \ldots, n-1\}$ represents the maximum number of connections for player i. If $M_i = N \setminus \{i\}$, player i can offer a connection to any player; in particular, if $a_i = n-1$, player i can have any number of connections. The result of the first stage is a network $g(b_1^1, \ldots, b_n^1)$ consisting of links (connections) ij such that $b_{ij}^1 = b_{ji}^1 = 1$. For brevity, denote $g(b_1^1, \ldots, b_n^1)$ by g. Define the neighbors of player i in network g as elements of the set $N_i(g) = \{j \in N \setminus \{i\} : ij \in g\}$ or simply N_i. After the network formation stage z_1, players proceed to the second stage z_2.

At second stage $z_2(g)$ which depends upon a network chosen at the first stage, network neighbors play pairwise simultaneous bimatrix games $\{\gamma_{ij}\}$. Namely, let $i \in N, j \in N_i$, then at the second stage, player i plays with his neighbor j a bimatrix game γ_{ij} with non-negative payoff matrices $A_{ij} = [a_{p\ell}^{ij}]_{p=1,\ldots,m; \ell=1,\ldots,k}$ and $B_{ij} = [b_{p\ell}^{ij}]_{p=1,\ldots,m; \ell=1,\ldots,k}$ for players i and j, respectively.

After receiving payoffs in these bimatrix games, the game ends. In other words, we have a two-stage game Γ which is a special case of a multistage non-zero-sum game. Adapting the definition of a strategy to this case, a *strategy* of player $i \in N$ will be a rule which assigns a set of his neighbors at first stage b_i^1, and a behavior b_i^2 in each of the bimatrix games at the second stage of the game taking into account a network formed at the first stage. Denote the strategy of player $i \in N$ in two-stage game Γ by $u_i = (b_i^1, b_i^2)$. Let (z_1, z_2) be a trajectory realized under the strategy profile $u = (u_1(\cdot), \ldots, u_n(\cdot))$ in Γ. Define the payoff of player i as $h_i(z_2)$ which is the sum of player i's payoffs in all bimatrix games with his neighbors when b_i^2, b_j^2, $j \in N_i$ are chosen. Then player i's payoff function in Γ starting at z_1 is defined as $K_i(z_1; u_i(\cdot), \ldots, u_n(\cdot)) = h_i(z_2)$.

The rest of the paper will be devoted to a cooperative version of the two-stage game Γ.

2.2.1 Cooperation at the Second Stage of the Game

A game Γ_{z_2} denoting a subgame of game Γ which starts at the second stage z_2 can be considered in cooperative form. In this case, we define characteristic function $v(z_2; S)$ for any subset (coalition) $S \subset N$ as the maxmin value of a two-person zero-sum game between coalition S and its complement $N \setminus S$ constructed with the use of game Γ_{z_2}. The superadditivity of the characteristic function follows from its definition. Denote the maxmin value of player i (j) in game γ_{ij} with his neighbor j (i) as

$$w_{ij} = \max_p \min_\ell a^{ij}_{p\ell}, \quad p = 1, \ldots, m, \ \ell = 1, \ldots, k,$$

$$w_{ji} = \max_\ell \min_p b^{ji}_{p\ell}, \quad p = 1, \ldots, m, \ \ell = 1, \ldots, k.$$

Following [2], for any $S \subseteq N$, the characteristic function $v(z_2; S)$ is given by:

$$v(z_2; S) = \begin{cases} \dfrac{1}{2} \displaystyle\sum_{i \in N} \sum_{j \in N_i} \max_{p,\ell}(a^{ij}_{p\ell} + b^{ji}_{p\ell}), & S = N, \\[2ex] \dfrac{1}{2} \displaystyle\sum_{i \in S} \sum_{j \in N_i \cap S} \max_{p,\ell}(a^{ij}_{p\ell} + b^{ji}_{p\ell}) + \sum_{i \in S} \sum_{k \in N_i \setminus S} w_{ik}, & S \subset N, \quad |S| > 2, \\[2ex] \displaystyle\max_{p,\ell}(a^{ij}_{p\ell} + b^{ji}_{p\ell}) + \sum_{r \in N_i \setminus \{j\}} w_{ir} + \sum_{q \in N_j \setminus \{i\}} w_{jq}, & S = \{i, j\}, \\[2ex] \displaystyle\sum_{j \in N_i} w_{ij}, & S = \{i\}, \\[2ex] 0, & S = \varnothing. \end{cases}$$

$$\tag{2.1}$$

2.2.2 Cooperation at Both Stages of the Game

Consider a cooperative form of two-stage game Γ. Suppose that all players choose strategies $\bar{u} = (\bar{u}_1, \ldots, \bar{u}_n)$ which maximize their joint payoff in game Γ, i.e.,

$$\sum_{i \in N} K_i(z_1; \bar{u}_1, \ldots, \bar{u}_n) = \max_u \sum_{i \in N} K_i(z_1; u_1, \ldots, u_n)$$

The strategy profile $\bar{u} = (\bar{u}_1, \ldots, \bar{u}_n)$ is called the *cooperative strategy profile*, and the corresponding trajectory $(\bar{z}_1, \bar{z}_2)$ is the *cooperative trajectory*.

As before for coalition $S \subseteq N$, we define characteristic function $v(\bar{z}_1; S)$ as the maxmin value of a two-person two-stage zero-sum game between coalition S and its complement, where the payoff of S is the sum of players' payoffs from S, and the strategy of S is an element of the Cartesian product of sets of players' strategies belonging to S. Since players' payoffs are non-negative, for player $N \setminus S$, the best behavior to follow is to have no connections with S. Hence we get

$$
v(\bar{z}_1; S) = \begin{cases} v(\bar{z}_2; N), & S = N, \\[2mm] \dfrac{1}{2} \sum_{i \in S} \sum_{j \in N_i \cap S} v(\bar{z}_1; \{i, j\}), & S \subset N, \ |S| > 2, \\[2mm] \max_{p, \ell}(a_{p\ell}^{ij} + b_{p\ell}^{ji}), & S = \{i, j\}, \\[2mm] 0, & |S| = 1, \text{ or } S = \varnothing. \end{cases} \tag{2.2}
$$

2.2.3 The Shapley Value and Time Consistency

Given a characteristic function $v(\bar{z}_t; \cdot), t = 1, 2$, we define an imputation as a vector $\xi[v(\bar{z}_t)] = (\xi_1[v(\bar{z}_t)], \ldots, \xi_n[v(\bar{z}_t)])$ which is (i) efficient, i.e., $\sum_{i \in N} \xi_i[v(\bar{z}_t)] = v(\bar{z}_t; N)$ and (ii) individually rational, i.e., $\xi_i[v(\bar{z}_t)] \geqslant v(\bar{z}_t; \{i\})$ for all $i \in N$. Denote the set of all imputations (an imputation set) in game Γ by $I(v(\bar{z}_t))$. As an imputation we consider the Shapley value [12] denoted by $\varphi[v(\bar{z}_t)] = (\varphi_1[v(\bar{z}_t)], \ldots, \varphi_n[v(\bar{z}_t)])$ where

$$
\varphi_i[v(\bar{z}_t)] = \sum_{S \subseteq N, i \in S} \frac{(|S| - 1)!(n - |S|)!}{n!} [v(\bar{z}_t; S) - v(\bar{z}_t; S \setminus \{i\})], \quad i \in N. \tag{2.3}
$$

Before the start of game Γ, players agree on choosing cooperative trajectory $(\bar{z}_1, \bar{z}_2)$, i.e., the trajectory that yields the maximum joint payoff $v(\bar{z}_1; N)$, and we suppose that players allocate this payoff according to the Shapley value. This means that in Γ each player $i \in N$ expects his payoff to be equal to $\varphi_i[v(\bar{z}_1)]$. If players recalculate the Shapley value after the network formation stage (at the second stage), it turns out that the recalculated Shapley value $\varphi[v(\bar{z}_2)]$ differs from the previous one. This may lead to a violation of the cooperative agreement because some players may refuse to use their cooperative strategies. We say that the Shapley value as an allocation in the two-stage game is *time consistent* if $\varphi[v(\bar{z}_1)] = \varphi[v(\bar{z}_2)]$ (as players do not receive payoffs at the network formation stage), otherwise we call the Shapley value *time inconsistent*. In the former case, players follow the cooperative agreement not expecting that someone violate it. In the latter case, to prevent players from violating the cooperative agreement, we use an *imputation distribution procedure* (IDP) $\beta = \{\beta_i^1, \beta_i^2\}_{i \in N}$ (first introduced in [8]) for the Shapley value $\varphi[v(\bar{z}_1)]$ which decomposes it over two stages of the game Γ: $\varphi_i[v(\bar{z}_1)] = \beta_i^1 + \beta_i^2$ for each $i \in N$. Here β_i^1 can be interpreted as a stage payment to player i at the network formation stage, and β_i^2 is his payment at the second stage of the game under the cooperative agreement. We say that the IDP β of the Shapley value $\varphi[v(\bar{z}_1)]$ is a *time-consistent IDP* [10, 11] when it is given by:

$$\beta_i^1 = \varphi_i[v(\bar{z}_1)] - \varphi_i[v(\bar{z}_2)], \qquad \beta_i^2 = \varphi_i[v(\bar{z}_2)], \qquad i \in N. \qquad (2.4)$$

Introducing the time-consistent IDP β (2.4) of the Shapley value $\varphi[v(\bar{z}_1)]$, players can be sure that no one violates the cooperative agreement, hence it will be realized in the game and player $i \in N$ gets $\varphi_i[v(\bar{z}_1)]$ as his cooperative payoff.

2.3 The Shapley Value for a Star

Since the calculation of the Shapley value $\varphi[v(\bar{z}_t)]$, $t = 1, 2$, is a difficult task for a large number of players in an arbitrary network, we simplify formula (2.3) for a network of a special type—a star. Within this section we suppose the following. Let $M_i = N \setminus \{i\}$ for $i \in N$ and $a_1 = n - 1$, $a_i = 1$, $i \neq 1$. Further let $\max_{j \in N} w_{ij} = w_{i1}$. Then in order to maximize the joint payoff, players should choose the following behaviors at the first stage of the game: $b_1^1 = (0, 1, \ldots, 1)$ for player 1, and $b_i^1 = (1, 0, \ldots, 0)$ for player $i \neq 1$. These behaviors form a star-network at this stage (see Fig. 2.1). In the star-network, $|N_1| = n - 1$ and $|N_i| = 1$, $i \neq 1$.

For a star-network, the characteristic function is calculated using a specific structure of the network. The network has central symmetry which suggests that formula (2.3) can be simplified. Let $m_{ij} = \max_{p,\ell}(a_{p\ell}^{ij} + b_{p\ell}^{ji})$. Substituting the adopted notation, as well as (2.1), (2.2), into (2.3), we obtain the following expression for the components of the Shapley value:

$$\varphi_i[v(\bar{z}_t)] = \begin{cases} \dfrac{1}{2}\left[v(\bar{z}_t; \{1\}) + \displaystyle\sum_{j \neq 1}\left(m_{1j} - v(\bar{z}_t; \{j\})\right)\right], & i = 1, \\[4mm] \dfrac{1}{2}[m_{1i} + v(\bar{z}_t; \{i\}) - w_{1i}], & i \neq 1. \end{cases} \qquad (2.5)$$

Fig. 2.1 A star with n players

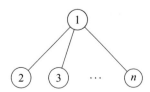

2.3.1 Two Examples

Two examples below demonstrate that the Shapley value being an allocation in a cooperative two-stage game with pairwise interactions can be both time consistent and time inconsistent. The first example shows the time consistency of the Shapley value.

Example 1 (Prisoner's Dilemma) Consider the case, when n players play the same game γ with their neighbors, i.e., $A_{ij} = A$, $B_{ij} = B$ for all $i \in N$, $j \in N_i$ where

$$A = B^T = \begin{pmatrix} b & 0 \\ a+b & a \end{pmatrix}, \quad 0 < a < b.$$

To find the Shapley value $\varphi[v(\bar{z}_2)]$, we first determine characteristic function $v(\bar{z}_2; S)$ for all $S \subseteq N$. Following (2.1), we obtain

$$v(\bar{z}_2; S) = \begin{cases} 2b(n-1), & S = N, \\ 2b(|S|-1) + (n-|S|)a, & S \subset N, \ 1 \in S, \\ |S|a, & S \subset N, \ 1 \notin S, \\ 0, & S = \varnothing. \end{cases}$$

Using the formula for the Shapley value (2.5) adapted to a star and noting that the Shapley value is an efficient allocation satisfying the property of symmetry and that $m_{1j} = 2b$ for any $j \in N_1$, we obtain

$$\varphi_1[v(\bar{z}_2)] = \frac{1}{2}[(n-1)a + (n-1)(2b-a)] = b(n-1),$$

$$\varphi_i[v(\bar{z}_2)] = \frac{v(\bar{z}_2; N) - \varphi_1[v(\bar{z}_2)]}{n-1} = b, \quad i \neq 1.$$

Similarly, to find the Shapley value $\varphi[v(\bar{z}_1)]$, we determine characteristic function $v(\bar{z}_1; S)$ for all $S \subseteq N$. Following (2.2), we have

$$v(\bar{z}_1; S) = \begin{cases} 2b(n-1), & S = N, \\ 2b(|S|-1), & S \subset N, \ 1 \in S, \\ 0, & S \subset N, \ 1 \notin S \quad \text{or} \quad S = \varnothing. \end{cases}$$

Again, using the formula for the Shapley value (2.5) adapted to a star, the Shapley value $\varphi[v(\bar{z}_1)]$ is given by

$$\varphi_1[v(\bar{z}_1)] = \frac{1}{2}[2b(n-1)] = b(n-1),$$

$$\varphi_i[v(\bar{z}_1)] = \frac{v(\bar{z}_1; N) - \varphi_1[v(\bar{z}_1)]}{n-1} = b, \quad i \neq 1.$$

Fig. 2.2 A star with four
players

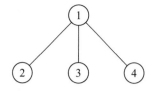

Comparing $\varphi[v(\bar{z}_1)]$ and $\varphi[v(\bar{z}_2)]$, we note they coincide and hence the Shapley value is time consistent.

In the next example we demonstrate the time inconsistency of the Shapley value.

Example 2 Consider a numerical example with $N = \{1, 2, 3, 4\}$ in which players form a star-network under a cooperative agreement (see Fig. 2.2). Let simultaneous bimatrix games γ_{12}, γ_{13}, and γ_{14} be defined by means of the following payoff matrices of players:

$$(A_{12}, B_{12}) = \begin{pmatrix} (2, 2) & (3, 0) \\ (5, 1) & (1, 2) \end{pmatrix}, \qquad (A_{13}, B_{13}) = \begin{pmatrix} (3, 1) & (4, 2) \\ (6, 2) & (2, 3) \end{pmatrix},$$

$$(A_{14}, B_{14}) = \begin{pmatrix} (1, 3) & (3, 2) \\ (6, 6) & (4, 1) \end{pmatrix}.$$

To compute the Shapley values $\varphi[v(\bar{z}_1)]$ and $\varphi[v(\bar{z}_2)]$, we use the corresponding formulas (2.1), (2.2) for characteristic functions $v(\bar{z}_2; \cdot)$ and $v(\bar{z}_1; \cdot)$, respectively, and the simplified formula (2.5). Hence we get

$$w_{12} = 2, \quad w_{13} = 3, \quad w_{14} = 4,$$
$$w_{21} = 1, \quad w_{31} = 2, \quad w_{41} = 3,$$
$$m_{12} = 6, \quad m_{13} = 8, \quad m_{14} = 12,$$

and therefore

$$v(\bar{z}_2; \{1\}) = 9, \quad v(\bar{z}_2; \{2\}) = 1, \quad v(\bar{z}_2; \{3\}) = 2, \quad v(\bar{z}_2; \{4\}) = 3,$$
$$v(\bar{z}_1; \{1\}) = 0, \quad v(\bar{z}_1; \{2\}) = 0, \quad v(\bar{z}_1; \{3\}) = 0, \quad v(\bar{z}_1; \{4\}) = 0,$$
$$v(\bar{z}_1; N) = 26, \quad v(\bar{z}_2; N) = 26.$$

Thus the Shapley values are given by

$$\varphi[v(\bar{z}_1)] = (13, 3, 4, 6), \qquad \varphi[v(\bar{z}_2)] = (29/2, 5/2, 7/2, 11/2).$$

We observe that the Shapley value $\varphi[v(\bar{z}_1)]$ in the two-stage game differs from the Shapley value $\varphi[v(\bar{z}_2)]$ in the one-stage game starting at the second stage. This means the time inconsistency of the Shapley value. Since $\varphi_2[v(\bar{z}_2)] = 5/2 < \varphi_2[v(\bar{z}_1)] = 3$, player 2 can break the cooperative agreement as his payoff can get less (here we recall that players do not receive payoffs at the network formation stage). Similar holds for player 3: $\varphi_3[v(\bar{z}_2)] = 7/2 < \varphi_3[v(\bar{z}_1)] = 4$ and player 4: $\varphi_4[v(\bar{z}_2)] = 11/2 < \varphi_4[v(\bar{z}_1)] = 6$. However introducing the time-consistent IDP of the Shapley value $\varphi[v(\bar{z}_1)]$ over two stages determined by formula (2.4), we obtain

$$\beta_1^1 = -3/2, \quad \beta_2^1 = 1/2, \quad \beta_3^1 = 1/2, \quad \beta_4^1 = 1/2,$$
$$\beta_1^2 = 29/2, \quad \beta_2^2 = 5/2, \quad \beta_3^2 = 7/2, \quad \beta_4^2 = 11/2,$$

and therefore cooperation will be sustainable. Thus receiving β_i^1 at the first stage and β_i^2 at the second stage, player $i \in N$ will get $\varphi_i[v(\bar{z}_1)]$ in two stages which is exactly player i's cooperative payoff prescribed by the Shapley value $\varphi[v(\bar{z}_1)]$.

2.4 Conclusion

In this paper, we studied a two-stage network game for a special type of pairwise interactions between players. This gave us the possibility of getting analytic expressions for characteristic functions in this game. As a solution of the game under consideration, we took the Shapley value and found its analytic form for a star-network. The special structure of the network game gives us the possibility of the implementation of other cooperative solutions what enriches the scope of application. The time inconsistency of the Shapley value in the two-stage game with pairwise interactions was demonstrated, and time-consistent IDP-based payoffs were introduced to deal with time inconsistency.

Acknowledgements This research was supported by the Russian Science Foundation (grant No. 17-11-01079).

References

1. Acemoglu, D., Ozdaglar, A., ParandehGheibi, A.: A Spread of (mis)information in social networks. Game Econ. Behav. **70**(2), 194–227 (2010)
2. Bulgakova, M.A., Petrosyan, L.A.: About strongly time-consistency of core in the network game with pairwise interactions. In: Proceedings of 2016 International Conference "Stability and Oscillations of Nonlinear Control Systems", pp. 157–160 (2016)
3. Dyer, M., Mohanaraj, V.: Pairwise-interaction games. In: Aceto, L., Henzinger, M., Sgall, J. (eds.) Automata, Languages and Programming. ICALP 2011. Lecture Notes in Computer Science, vol. 6755, pp. 159–170 (2011)
4. Dziubiński, M., Goyal, S.: Network design and defence. Game Econ. Behav. **79**, 30–43 (2013)
5. Hernández, P., Muñoz-Herrera, M., Sánchez, Á.: Heterogeneous network games: conflicting preference. Game Econ. Behav. **79**, 56–66 (2013)
6. König, M.D., Battiston, S., Napoletano, M., Schweitzer, F.: The efficiency and stability of R&D networks. Game Econ. Behav. **75**(2), 694–713 (2012)

7. Kuzyutin, D., Nikitina, M.: Time consistent cooperative solutions for multistage games with vector payoffs. Oper. Res. Lett. **45**(3), 269–274 (2017)
8. Petrosyan, L.A., Danilov, N.N.: Stability of solutions of non-zero-sum game with transferable payoffs. Vestn. Leningr. Univ. Ser 1. Mat. Mekh. Astron. **19**, 52–59 (1979)
9. Petrosyan, L.A., Sedakov, A.A.: Multistage network games with perfect information. Autom. Remote Control **75**(8), 1532–1540 (2014)
10. Petrosyan, L.A., Sedakov, A.A.: The subgame-consistent Shapley value for dynamic network games with shock. Dyn. Games Appl. **6**(4), 520–537 (2016)
11. Petrosyan, L.A., Sedakov, A.A., Bochkarev, A.O.: Two-stage network games. Autom. Remote Control **77**(10), 1855–1866 (2016)
12. Shapley, L.: A value for N-person games. In: Kuhn, H.W., Tucker, A.W. (eds). Contributions to the Theory of Games II, pp. 307–317. Princeton University Press, Princeton (1953)

Chapter 3
Routing on a Ring Network

Ramya Burra, Chandramani Singh, Joy Kuri, and Eitan Altman

3.1 Introduction

Routing problems arise in networks in which common resources are shared by a group of users. Examples of such scenario include flow routing in communication networks, traffic routing in transportation networks, flow of work in manufacturing plants, etc. Each user incurs a certain cost (e.g., delay) at each link on its route, where the cost depends on the flows through the link. The routing problems, when handled by a centralized controller, aim to optimize the aggregate cost of all the users, e.g., average network delay. However, a centralized solution may not be viable for several reasons. For instance, a very large network and its time varying attributes (e.g., traffic and link states in a communication network) could lead to excessive communication overhead for solving the problem centrally. In other cases, the very premise of the network may be such that local administrators control different portions of the network, e.g., different depots controlling different parts of a transportation network. In either case, distributed controllers may compete to maximize individual, and often conflicting, performance measures. It is imperative to assess the performance of distributed control, especially how far it is from the global optimal.

R. Burra (✉) · C. Singh · J. Kuri
Department of ESE, Indian Institute of Science, Bangalore, India
e-mail: burra@iisc.ac.in; chandra@iisc.ac.in; kuri@iisc.ac.in

E. Altman
Univ Côte d'Azur, INRIA, Sophia Antipolis, Méditerranée, France

LINCS, Paris, France
e-mail: Eitan.Altman@inria.fr

© Springer Nature Switzerland AG 2019
J. B. Song et al. (eds.), *Game Theory for Networking Applications*,
EAI/Springer Innovations in Communication and Computing,
https://doi.org/10.1007/978-3-319-93058-9_3

Fig. 3.1 A ring network with
$K = 2$. For example, the
users at node 1 can use paths
$(1, 0)$, $(1, 2, 0)$, and
$(1, 2, 3, 0)$

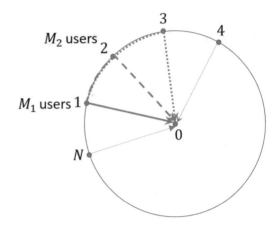

Distributed control of routing has been widely modelled as noncooperative games among self-interested decision makers. Nash equilibria of the games (Wardrop equilibria in case of nonatomic games) characterize the system-wide flow configuration resulting from such distributed control. Wardrop [10] introduced Wardrop equilibrium in the context of transportation networks, and Dafermos and Sparrow [4] showed that it can be characterized as a solution of a standard network optimization problem. Orda et al. [8] showed existence and uniqueness of Nash equilibrium in routing games under various assumptions on the cost function. They also showed a few interesting monotonicity properties of the Nash equilibria. Cominetti et al. [3] computed the worst-case inefficiency of Nash equilibria and also provided a pricing mechanism that reduces the worst-case inefficiency. Altman et al. [1] considered a class of polynomial link cost functions and showed that these lead to predictable and efficient Nash equilibria. Hanwal et al. [5] studied routing over time and studied a stochastic game resulting from random arrival of traffic.

We study a routing problem on a ring network in which users' traffic originate at nodes on the ring and are destined to a common node at the center (see Fig. 3.1). Each user can use the direct link from its node to the center and also a certain number of paths through the adjacent nodes, to transport its traffic. The users incur two costs: (i) The cost of using a link between a node at the ring and the center, (ii) the cost of redirecting the traffic through adjacent nodes. The number of users attached to the node can be random. We characterize Nash equilibria of such routing games.

Scheduling problems are a class of resource allocation problems in which resources are shared over time. In these problems, unlike simultaneous action routing problems, each user may see the system state that results from its predecessor's actions. However, if we assume that such information is not available to the users, our framework can also be used to analyze certain scheduling (or temporal routing) problem.

In Sect. 3.2, we formally introduce our general framework and also illustrate how it can be used to model several problems arising in communication networks, transportation networks, etc. In subsequent sections, we analyze special cases of this framework. Following is a brief outline of our contribution

1. In Sect. 3.3, we show that routing games with only one and two hop paths and linear costs are potential games. We also give explicit expressions of Nash equilibrium flows for networks with any generic cost function and symmetric loads.
2. In Sect. 3.4 we consider networks with random loads and linear routing costs. We give explicit characterization of Nash equilibria for two cases: (i) General load distribution and one and two hop paths, (ii) Bernoulli distributed loads.

The omitted proofs can be found in our technical report [2].

3.2 System Model

Let us consider a ring network with N nodes and M_n users at each node $n \in [N] := \{1, \ldots, N\}$. Let us assume that the ith user at node n has a flow requirement ϕ_n^i to be sent to the center. Let $c(z)$ represent the cost per unit of flow at any link where z is the aggregate traffic through this link. Throughout we assume that $c(\cdot)$ is positive, strictly increasing and convex. We assume that each user can use the direct link to the center and the K other links through K adjacent nodes in the clockwise direction. For example, any user at node n can use links $(n, 0), (n+1, 0), \ldots, (n+K, 0)$.[1] We also assume that a user at node n incurs kd extra per unit flow cost for any flow that it routes through link $(n+k, 0)$. Note that we assume no cost for using the links along the ring.

For each $n \in [N]$, $i \in [M_n]$, $l \in [n, n+K]$, let x_{nl}^i be the flow of ith user at node n that is routed through link l. We let x_n^i denote the flow configuration of the ith user at node n, x denote the network flow configuration, and x_l denote the total flow through link l; $x_n^i = (x_{nl}^i, l \in [n, n+K])$, $x = (x_n^i, n \in [N], i \in [M_n])$ and $x_l = \sum_{n \in [l-K, l]} \sum_{i \in [M_n]} x_{nl}^i$. Then the total cost of ith user is

$$C_n^i(x) = \sum_{l \in [n, n+K]} x_{nl}^i (c(x_l) + (l - n)d), \tag{3.1}$$

and the aggregate network cost is $C(x) = \sum_{n \in [N]} \sum_{i \in [M_n]} C_n^i(x)$. Note that the flows must satisfy

$$\sum_{l \in [n, n+K]} x_{nl}^i = \phi_n^i \tag{3.2}$$

for all $i \in [M_n]$, $n \in [N]$ in addition to nonnegativity constraints.

We now illustrate how this framework can model a variety of routing and scheduling problems.

---

[1] Clearly, the addition here is modulo N.

1. We can think of this framework as modeling routing in a transportation network in a city. The ring and the center represent a ring road and the city center, respectively. We have sets of vehicles starting from various entry points, represented as nodes on the ring, all destined to the city center. The costs here represent latency. We assume that the ring road has large enough capacity to render the latency along it independent of the load. On the other hand, latency on the roads joining the ring to the center is traffic dependent. Each node has a set of depots, each controlling routing of a subset of vehicles starting at this node.
2. We can use this framework to model load balancing in distributed computer systems [6].
3. We can also use this framework to model scheduling of charging of electric vehicles at a charging station. Here, the nodes represent time slots and players represent vehicles. The per unit charging cost in a slot depends on the charge drawn in that slot. Each vehicle can wait up to K slots to be charged. We assume that the vehicles do not know pending charge from the earlier vehicles when making scheduling decision.

3.3 Deterministic Loads

Nash Equilibrium A flow configuration x is a Nash equilibrium if, for all $i \in [M_n], n \in [N]$,

$$C_n^i(x) = \min_{y_n^i} C_n^i(y_n^i, x \setminus x_n^i) \tag{3.3}$$

subject to (3.2) and nonnegative constraints. Under our assumptions on $c(\cdot)$, the routing game is a *convex game* [9]. Existence and uniqueness of the Nash equilibrium then follows from [8]. It follows that the equilibrium is characterized by the following Kuhn-Tucker conditions(using cost from Eq. (3.1)): for every $i \in [M_n]$ there exists a Lagrange multiplier λ_n^i such that, for every link $l \in [n, n + K]$,

$$c(x_l) + (l - n)d + x_{nl}^i c'(x_l) \geq \lambda_n^i \tag{3.4}$$

with equality if $x_{nl}^i > 0$. From this,

$$\lambda_n^i = \frac{\sum_{l:x_{nl}^i>0} \frac{c(x_l)+(l-n)d}{c'(x_l)} + \phi_n^i}{\sum_{l:x_{nl}^i>0} \frac{1}{c'(x_l)}}, \quad \text{for all } i \in [M_n], n \in [N].$$

We observe that the equilibrium flow configuration is the solution of the following system of equations.

$$x_{nj}^i = \max\left\{\frac{1}{c'(x_j)}\frac{\sum_{l:x_{nl}^i>0}\frac{c(x_l)-c(x_j)+(l-j)d}{c'(x_l)}+\phi_n^i}{\sum_{l:x_{nl}^i>0}\frac{1}{c'(x_l)}}, 0\right\}, \tag{3.5}$$

for all $i \in [M_n]$, $j \in [n, n+K]$, $n \in [N]$.

We can elegantly obtain Nash equilibria in special cases. In the following two subsections we consider two such cases, the first allowing only one-hop and two-hop paths to the center and linear costs, and the second having same number of users, all with identical requirements, at all the nodes.

3.3.1 Maximum Two Hops and Linear Costs ($K = 1$ and $c(x) = x$)

Here, using $x_{nn}^i + x_{n(n+1)}^i = \phi_n^i$, from (3.5),

$$x_{nn}^i = \left[\frac{c(x_{n+1}) - c(x_n) + d + c'(x_{n+1})\phi_n^i}{c'(x_n) + c'(x_{n+1})}\right]_0^{\phi_n^i}$$

for all $i \in [M_n]$, $n \in [N]$.[2] For linear costs, substituting $c(x) = x$ for all x,

$$x_{nn}^i = \left[\frac{x_{n+1} - x_n + d + \phi_n^i}{2}\right]_0^{\phi_n^i}. \tag{3.6}$$

Further, using

$$x_n = \sum_{j\in[M_n]} x_{nn}^j + \sum_{j\in[M_{n-1}]} (\phi_{n-1}^j - x_{(n-1)(n-1)}^j)$$

$$\text{and } x_{n+1} = \sum_{j\in[M_n]} (\phi_n^j - x_{nn}^j) + \sum_{j\in[M_{n+1}]} x_{(n+1)(n+1)}^j,$$

$$x_{nn}^i = \left[\frac{2\phi_n^i + \sum_{j\in M_n\setminus i}(\phi_n^j - 2x_{nn}^j)}{4}\right.$$

$$\left.+\frac{\sum_{j\in M_{n+1}} x_{(n+1)(n+1)}^j - \sum_{j\in M_{n-1}}(\phi_{n-1}^j - x_{(n-1)(n-1)}^j) + d}{4}\right]_0^{\phi_n^i}. \tag{3.7}$$

---

[2] $[x]_a^b := \min\{\max\{x, a\}, b\}$.

Notice that flow configuration of ith user at node n is completely specified by x^i_{nn}. The above equation can be seen as the best response of this user.

Lemma 1 *If the players update according to (3.7), round-robin or random update processes converge to the Nash equilibrium.*

Proof The routing game under consideration is a potential game with potential function

$$V(x) = \frac{1}{2} \left(\sum_{n \in [N]} \left(x_n^2 + \sum_{i \in [M_n]} ((x^i_{nn})^2 + (x^i_{n(n+1)})^2 + 2x^i_{n(n+1)}d) \right) \right).$$

Hence it exhibits the improvement property and convergence as stated in the lemma [7]. □

3.3.2 Symmetric Loads ($M_n = M$ and $\phi^i_n = \phi$)

Here, we can restrict to symmetric flow configurations owing to symmetry of the problem. We can express any symmetric network flow configuration as a vector $\beta = (\beta_0, \beta_1 \ldots, \beta_K)$, $\sum_{j=0}^K \beta_j = \phi$ where $\beta_j := x^i_{n(n+j)}$ for all $i \in [M]$, $n \in [N]$ and $j \in [0, K]$.

Theorem 1 *If $M_n = M$ and $\phi^i_n = \phi$ for all $i \in [M_n]$, $n \in [N]$, then the unique Nash equilibrium is*

$$\beta_j = \begin{cases} \frac{\phi}{k^*+1} + \frac{(K^*-2j)d}{2c'(M\phi)} \ if\ l \in [n, n+K^*] \\ 0\ otherwise. \end{cases} \tag{3.8}$$

where $K^* = \min\{\max\{k : k(k+1) < \frac{2\phi c'(M\phi)}{d}\}, K\}$

Proof From the Karush-Kuhn-Tucker conditions for optimality of β (see (3.4)),

$$c(\beta_j + x^{-i}_{n(n+j)}) + jd + \beta_j c'(\beta_j + x^{-i}_{n(n+j)}) \geq \lambda \tag{3.9}$$

where $x^{-i}_{n(n+j)}$ is the total flow on link $(n+j, 0)$ except that of ith user at node n. Note that, for β to be a symmetric Nash equilibrium, $\beta_j + x^{-i}_{n(n+j)} = M\phi$. Hence (3.9) can be reduced to

$$c(M\phi) + jd + \beta_j c'(M\phi) \geq \lambda$$

with equality if $\beta_j > 0$. So, we see that

$$\beta_j = \max \left\{ \frac{1}{c'(M\phi)} (\lambda - c(M\phi) - jd), 0 \right\} \tag{3.10}$$

Note that β_j is decreasing in j. Let us assume that $\beta_k > 0$ for all $k \leq K'$ for some $K' \leq K$, and 0 otherwise. Then, using $\sum_{i=0}^{K'} \beta_i = \phi$ in (3.10),

$$\lambda(K') - c(M\phi) = \frac{\phi c'(M\phi)}{(K'+1)} + \frac{K'd}{2}, \tag{3.11}$$

where we write $\lambda(K')$ to indicate dependence of λ on K'. Substituting the above back in (3.10),

$$\beta_j = \frac{\phi}{1+K'} + \frac{d(K'-2j)}{2c'(M\phi)}, \, j \in [0, K']. \tag{3.12}$$

To complete the proof, we claim that K' equals K^* where

$$K^* = \min \left\{ \max \left\{ k : k(k+1) < \frac{2\phi c'(M\phi)}{d} \right\}, K \right\}.$$

Let us first argue that K' cannot exceed K^*. We only need to consider the case when $K^* < K$. In this case, from the definition of K^*, for any $K' > K^*$,

$$\frac{1}{K'+1} - \frac{dK'}{2\phi c'(M\phi)} \leq 0,$$

which contradicts the defining property of K' that $\beta_k > 0$ for all $k \leq K'$. This completes the argument. Now we argue that K' cannot be smaller than K^*, again by contradiction. Let $K' < K^*$. Then, from (3.11),

$$\lambda(K') - \lambda(K^*) = \frac{\phi c'(M\phi)(K^* - K')}{(K^*+1)(K'+1)} - \frac{(K^* - K')d}{2}$$

$$= \phi c'(M\phi)(K^* - K') \left\{ \frac{1}{(K^*+1)(K'+1)} - \frac{d}{2\phi c'(M\phi)} \right\} > 0,$$

where the inequality follows from definition of K^*. Hence, from (3.10),

$$\beta_{K^*} \geq \frac{1}{c'(M\phi)} (\lambda(K') - c(M\phi) - K^*d)$$

$$> \frac{1}{c'(M\phi)} (\lambda(K^*) - c(M\phi) - K^*d) > 0.$$

This contradicts $K' < K^*$ which would imply $\beta_{K^*} = 0$. □

Optimal Routing The optimal strategy of any user will be $\beta_0 = \phi$ and $\beta_j = 0$, $1 \leq j \leq K$.

3.4 Random Loads

We now consider the scenario where the numbers of users at various nodes, M_n, are i.i.d random variables with distribution $(p_1, p_2 \ldots, p_M)$. We assume that a user knows the number of collocated users but only knows the distribution of users at the other nodes. Throughout this section we restrict to equal flow requirements for all the users, i.e. $\phi_n^i = \phi$ for all $n \in [N], i \in [M_n]$, and linear per unit flow cost, i.e., $c(x) = x$. In the following we analyze two special cases of this routing problem, the first assuming the users can only use one-hop and two-hop paths to center, and the second having Bernoulli user distribution.

3.4.1 Maximum Two Hops (K = 1)

We consider symmetric flow configurations where all the users with equal number of collocated users adopt same flow configuration. We can then express the network flow configuration as a vector $\gamma = (\gamma(1), \gamma(2), \ldots)$ where $\gamma(m)$ represents the flow that a user with m collocated users redirects to its two-hop path. Let us define

$$\bar{P}_m = 1 - \sum_{l=m+1}^{M} \frac{l p_l}{l+1} \text{ and } Q_m = \sum_{l=0}^{m} l p_l,$$

for all $0 \le m \le M$.

Theorem 2 *The unique Nash equilibrium is given by*

$$\gamma(m) = \begin{cases} 0, & \text{if } 1 \le m \le m_\alpha \\ \frac{\phi}{2} - \frac{(d-\alpha)}{2(m+1)}, & \text{otherwise,} \end{cases}$$

$$\text{where } m_\alpha = \min\left\{ \min\left\{ m : \frac{d}{\phi \bar{P}_m} + \frac{Q_m}{\bar{P}_m} < m+2 \right\}, M \right\}$$

$$\text{and } \alpha = d - \frac{d}{\bar{P}_{m_\alpha}} - \frac{Q_{m_\alpha} \phi}{\bar{P}_{m_\alpha}}.$$

Proof Let us consider a user i with m collocated users. Let us fix the strategies of all other users in the network to $\gamma = (\gamma(1), \gamma(2), \ldots)$. Then the best response of user i, say $\gamma'(m)$, is the unique minimizer of the cost function

$$(\phi - \gamma'(m))((m-1)(\phi - \gamma(m)) + \phi - \gamma'(m) + \sum l p_l \gamma(l))$$

$$+ \gamma'(m)((m-1)\gamma(m) + \gamma'(m) + \sum_l l p_l (\phi - \gamma(l)) + d).$$

$\gamma'(m)$ must satisfy the following optimality criterion

$$-2(\phi - \gamma'(m)) - (m-1)(\phi - \gamma(m)) - \sum lp_l \gamma(l)$$
$$+ 2\gamma'(m) + (m-1)\gamma(m) + \sum lp_l(\phi - \gamma(l)) + d \geq 0$$

with equality if $\gamma'(m) > 0$. For γ to be a symmetric Nash equilibrium, setting $\gamma'(m) = \gamma(m)$ in the above inequality,

$$-(m+1)\phi + 2(m+1)\gamma(m) + \sum lp_l \phi - 2\sum lp_l \gamma(l) + d \geq 0$$

yielding

$$\gamma(m) = \max \left\{ \frac{\phi}{2} + \frac{2\sum lp_l \gamma(l) - \phi \sum lp_l - d}{2(m+1)}, 0 \right\}$$

Clearly, the above should hold for all $m \in \{0, 1, \ldots, M\}$. Setting,

$$\alpha = 2\sum lp_l \gamma(l) - \phi \sum lp_l, \tag{3.13}$$

$$\text{and } m_\alpha = \left\lfloor \frac{d-\alpha}{\phi} - 1 \right\rfloor, \tag{3.14}$$

we get

$$\gamma(m) = \begin{cases} \frac{\phi}{2} + \frac{\alpha-d}{2(m+1)}, & \text{if } m > m_\alpha \\ 0, & \text{otherwise.} \end{cases} \tag{3.15}$$

We now show how to obtain α and m_α. From (3.15),

$$mp_m(2\gamma(m) - \phi) = \begin{cases} \frac{(\alpha-d)mp_m}{m+1}, & \text{if } m > m_\alpha \\ -mp_m\phi, & \text{otherwise.} \end{cases}$$

Using this in (3.13),

$$\alpha = \frac{-d \sum_{m>m_\alpha} \frac{mp_m}{m+1} - \phi \sum_{m=0}^{m_\alpha} mp_m}{\left(1 - \sum_{m>m_\alpha} \frac{mp_m}{m+1}\right)}$$
$$= d - \frac{d}{\bar{P}_{m_\alpha}} - \frac{Q_{m_\alpha}\phi}{\bar{P}_{m_\alpha}},$$

and hence, from (3.14),

$$m_\alpha = \left\lfloor \frac{d}{\phi \bar{P}_{m_\alpha}} + \frac{Q_{m_\alpha}}{\bar{P}_{m_\alpha}} - 1 \right\rfloor$$

Let us now turn to the expression of m_α in the statement of the theorem. Clearly, $m_\alpha > \frac{d}{\phi \bar{P}_{m_\alpha}} + \frac{Q_{m_\alpha}}{\bar{P}_{m_\alpha}} - 2$. Also,

$$\frac{d}{\phi \bar{P}_{m_\alpha - 1}} + \frac{Q_{m_\alpha - 1}}{\bar{P}_{m_\alpha - 1}} \geq m_\alpha + 1,$$

implying

$$\frac{d}{\phi \bar{P}_{m_\alpha}} + \frac{Q_{m_\alpha}}{\bar{P}_{m_\alpha}} \geq m_\alpha + 1,$$

$$\text{or,}\ \frac{d}{\phi \bar{P}_{m_\alpha}} + \frac{Q_{m_\alpha}}{\bar{P}_{m_\alpha}} - 1 \geq m_\alpha.$$

So, the two expressions of m_α are equivalent, and $\gamma(m)$s in the statement of the theorem indeed constitute a Nash equilibrium. Also note that existence of an optimal γ ensures existence of at least one (α, m_α) pair satisfying (3.13)–(3.14). It remains to establish uniqueness of (α, m_α) pair satisfying (3.13)–(3.14). We do this in Appendix.   □

Optimal Routing The expected total routing cost will be N times the sum of expected routing costs on links $(n - 1, n)$ and $(n, 0)$ for an arbitrary n. In the following, we optimize the latter to get the optimal flow configuration.

Theorem 3 *The unique optimal flow configuration is given by*

$$\gamma(m) = \begin{cases} 0, & \text{if } 0 \leq m \leq m_{\bar{\alpha}} \\ \frac{\phi}{2} - \frac{(d - \bar{\alpha})}{4m}, & \text{otherwise,} \end{cases}$$

$$\text{where } m_{\bar{\alpha}} = \min \left\{ \min \left\{ m : \frac{d}{2\phi P_m} + \frac{Q_m}{P_m} < m + 1 \right\}, M \right\}$$

$$\text{and } \bar{\alpha} = d - \frac{d}{P_{m_{\bar{\alpha}}}} - \frac{2Q_{m_{\bar{\alpha}}}\phi}{P_{m_{\bar{\alpha}}}}.$$

3.4.2 Bernoulli Loads ($p_0 + p_1 = 1$)

We again focus on only symmetric flow configuration. As in Sect. 3.3.2, we let $x^i_{nl} = \beta_{n-l}$ for all $i \in [M]$, $n \in [N]$ and $l \in [n, n + K]$.

Theorem 4 *The unique Nash equilibrium is given by*

$$
\beta_i = \begin{cases} \frac{\phi}{K^*+1} + \frac{d(K^*-2i)}{2(2-p)}, & if\ 0 \le i \le K^* \\ 0, & otherwise, \end{cases}
$$

where $K^* = \min\{\max\{k : k(k+1) < \frac{2\phi(2-p)}{d}\}, K\}$.

Optimal Routing The expected total routing cost will be N times the sum of expected routing costs on links $(n-1, n)$ and $(n, 0)$ for an arbitrary n. In the following, we optimize the latter to get the optimal flow configuration.

Theorem 5 *The unique optimal flow configuration is given by*

$$
\beta_j = \begin{cases} \frac{\phi}{K^*+1} + \frac{d(K^*-2j)}{4(1-p)}, & if\ 0 \le i \le K^* \\ 0, & otherwise, \end{cases}
$$

where $K^* = \min\{\max\{k : k(k+1) < \frac{4(1-p)\phi}{d}\}, K\}$.

3.5 Conclusion and Future Work

We studied routing on a ring network. We studied both, non-cooperative games between competing users and network optimal routing. We considered several special cases of networks with deterministic and random loads. We provided characterization of Nash equilibria and optimal flow configuration in these cases (see Theorems 1–5).

Our future work entails extending this analysis to more general cases. We would like to study price of anarchy, and also pricing mechanisms (tolls) that induce optimality.

Acknowledgements The first and second authors acknowledge support from Visvesvaraya PhD Scheme and INSPIRE Faculty Research Grant (DSTO-1363), respectively. This work is also partly supported by CEFIPRA/Inria Grant IFC/DST-Inria-2016-01/448.

Appendix

We establish uniqueness via contradiction. Let (α, m_α) and $(\alpha', m_{\alpha'})$ be two pairs satisfying (3.13)–(3.14). We assume $m_{\alpha'} > m_\alpha$ without any loss of generality. Recall that $\alpha = d - \frac{d}{\bar{P}_{m_\alpha}} - \frac{\phi Q_{m_\alpha}}{\bar{P}_{m_\alpha}}$ and $\frac{d-\alpha}{\phi} < m_\alpha + 2$, implying

$$
\frac{d}{\phi} < (m_\alpha + 2)\bar{P}_{m_\alpha} - Q_{m_\alpha}. \tag{3.16}
$$

Similarly, $\alpha' = d - \frac{d}{P_{m_{\alpha'}}} - \frac{\phi Q_{m_{\alpha'}}}{P_{m_{\alpha'}}}$ and $\frac{d-\alpha'}{\phi} \geq m_{\alpha'} + 1$, implying

$$\frac{d}{\phi} \geq (m_{\alpha'} + 1)\bar{P}_{m_{\alpha'}} - Q_{m_{\alpha'}}. \tag{3.17}$$

We argue that $(m_{\alpha'} + 1)\bar{P}_{m_{\alpha'}} - Q_{m_{\alpha'}} \geq (m_\alpha + 2)\bar{P}_{m_\alpha} - Q_{m_\alpha}$, and hence both (3.16) and (3.17) cannot hold simultaneously. Indeed note that

$$(m_{\alpha'} + 1)\bar{P}_{m_{\alpha'}} - (m_\alpha + 2)\bar{P}_{m_\alpha}$$
$$= (m_{\alpha'} + 1)(\bar{P}_{m_{\alpha'}} - \bar{P}_{m_{\alpha'}-1}) + (m_{\alpha'} + 1)\bar{P}_{m_{\alpha'}-1} - (m_\alpha + 2)\bar{P}_{m_\alpha}$$
$$= m_{\alpha'} p_{m_{\alpha'}} + \sum_{m=m_\alpha+1}^{m_{\alpha'}-1} \{(m+2)\bar{P}_m - (m+1)\bar{P}_{m-1}\}$$
$$\geq m_{\alpha'} p_{m_{\alpha'}} + \sum_{m=m_\alpha+1}^{m_{\alpha'}-1} (m+1)(\bar{P}_m - \bar{P}_{m-1}) = \sum_{m=m_\alpha+1}^{m_{\alpha'}} m p_m = Q_{m_{\alpha'}} - Q_{m_\alpha}$$

This completes the argument.

References

1. Altman, E., Başar, T., Jiménez, T., Shimkin, N.: Competitive routing in networks with polynomial cost. IEEE Trans. Autom. Control **47**(1), 92–96 (2002)
2. Burra, R., Singh, C., Kuri, J., Altman, E.: Routing on a ring network (February 2018). http://chandramani.dese.iisc.ac.in/wp-content/uploads/sites/6/2018/02/ramya-etal18routing-ring-networks.pdf
3. Cominetti, R., Correa, J.R., Stier-Moses, N.E.: The impact of oligopolistic competition in networks. Oper. Res. **57**(6), 1421–1437 (2009)
4. Dafermos, S.C., Sparrow, F.T.: The traffic assignment problem for a general network. J. Res. Natl. Bur. Stand. Ser. B **73B**(2), 91–118 (1969)
5. Hanawal, M.K., Altman, E., El-Azouzi, R., Prabhu, B.J.: Spatio-temporal control for dynamic routing games. In: International Conference on Game Theory for Networks, Shanghai, China, pp. 205–220 (April 2011)
6. Kameda, H., Altman, E., Kozawa, T., Hosokawa, Y.: Braess-like paradoxes in distributed computer systems. IEEE Trans. Autom. Control **45**(9), 1687–1691 (2000)
7. Monderer, D., Shapley, L.S.: Potential games. Games Econ. Behav. **14**(1), 124–143 (1996)
8. Orda, A., Rom, R., Shimkin, N.: Competitive routing in multi-user communication networks. IEEE/ACM Trans. Networking **1**(5), 510–521 (1993)
9. Rosen, J.B.: Existence and uniqueness of equilibrium points for concave N-person games. Econometrica **33**(3), 520–534 (1965)
10. Wardrop, J.G.: Some theoretical aspects of road traffic research. Proc. Inst. Civ. Eng. **1**(3), 325–378 (1952)

Chapter 4
Performance of Dynamic Secure Routing Game

Ju Bin Song and Quanyan Zhu

4.1 Introduction

In distributed secondary networks, routing is inherently fragile and can easily be compromised by unknown attacks [3]. Consequently, it is imperative to design routing schemes that can enhance security of routing in distributed cognitive radio networks [4]. Moreover, the primary users (PUs) can affect the spectrum opportunities available for the secondary users (SUs), leading to dynamically changing network topology in multi-hop cognitive radio (CR) networks [6]. Therefore, the secure routing scheme needs to allow secondary users to learn their environment in a distributed and dynamic fashion and to yield optimal routing decisions that can defend against malicious attacks with minimum level of compromise in performances. The dynamic game framework has been applied to network formation problems in wireless mobile networks [9, 18, 21]. In our early works [19], users define their own hierarchies by their multi-stage exploration processes of neighboring nodes. In addition, the secure routing games capture the conflicting goals between the attackers and the users, whereas the network formation game models have considered interactions among competing agents of the same nature. Another challenge that we address here is the users' lack of knowledge of the action set of the attackers and their utility function in practice due to the distributed routing mechanism without centralized information.

J. B. Song (✉)
Department of Electronic Engineering, Kyung Hee University, Yongin, South Korea
e-mail: jsong@khu.ac.kr

Q. Zhu
Department of Electrical and Computer Engineering, NYU, Brooklyn, NY, USA
e-mail: quanyan.zhu@nyu.edu

© Springer Nature Switzerland AG 2019
J. B. Song et al. (eds.), *Game Theory for Networking Applications*,
EAI/Springer Innovations in Communication and Computing,
https://doi.org/10.1007/978-3-319-93058-9_4

Table 4.1 Summary of notations

Symbol	Meaning
$\mathcal{N}$	Set of secondary user nodes
$\mathcal{M}$	Set of primary user nodes
$\mathcal{J}$	Set of jammers
$s \in \mathcal{S}$	System state
$L_i, i \in \mathcal{N}$	Total number of explorations of secondary i till its destination
$\mathcal{A}_{l_i}$	Set of secondary nodes at stage l_i exploration of secondary i
$\mathcal{R}_j, j \in \mathcal{J}$	Set of secondary user nodes within the jamming range of jammer j
$\mathcal{J}_{l_i}$	Set of jammers encountered at stage l_i exploration of secondary user i
$\gamma_x^y, x, y \in \mathcal{N}$	Signal to interference plus noise ratio (SINR)
	between a generic node x and a generic node y
$I_x^y, x, y \in \mathcal{N}$	Interference between a generic node x and a generic node y
$\tau_x^y, x, y \in \mathcal{N}$	Packet delay between a generic node x and a generic node y
$(n_i, l_i) \in \mathcal{A}_{l_i}$	Node that routes data from secondary user i at stage l_i
$\mathcal{P}_i(l_i, l_i')$	Set of routing nodes between stage l_i and l_i'
$\mathcal{Q}_i(l_i, l_i')$	Routing path of secondary user i from stage l_i to l_i'
$u_{(n_i, l_i)}$	Stage utility of secondary user node (n_i, l_i) at stage $h = l_i$
$U_{(n_i, l_i)}$	Total utility along the path $\mathcal{Q}_{(n_i, l_i)}(l_i, L_i)$
$\mathrm{u}_{(n_i, l_i)}$	Expected stage utility of secondary user node (n_i, l_i)
$\mathbb{U}_{(n_i, l_i)}$	Expected total utility of secondary user node (n_i, l_i)
$\overline{U}_{(n_i, l_i)}$	Upper value with source secondary user node (n_i, l_i)
$\underline{U}_{(n_i, l_i)}$	Lower value with source secondary user node (n_i, l_i)
$U_{(n_i, l_i)}^*$	Value of the zero-sum game with source secondary user node (n_i, l_i)

In this paper, we analyze the multi-stage saddle-point equilibrium strategies using backward induction and compute real-time strategies using distributed learning. In addition, we analyze the network performance of proposed dynamic securing routing game framework using Network Simulator 2 (NS-2). The rest of this paper is organized as follows. Section 4.2 presents related works on secure routing techniques in cognitive radio networks. In Sect. 4.3, the system model is described. The dynamic game-theoretic model is analyzed in Sect. 4.4. Section 4.5 presents the saddle-point in mixed strategies for dynamic secure routing game. In Sect. 4.6, simulation results are described. Finally, conclusions are drawn in Sect. 4.7.

We summarize some of the notations used in the paper in Table 4.1.

4.2 Related Work

In this section, we describe the related works on secure routing techniques against unknown jamming attacks in distributed cognitive radio networks. Since secondary users have limited opportunities to utilize idle spectrum resources left by primary users, the performance routing path for SUs is inevitably influenced by the state of PUs as well as the communication activities of other SUs. In addition, in an adversarial environment, the established routing path can be fragile to jamming attacks. It is desirable that the implemented routing protocol establishes new path in event of unanticipated adversarial attacks and allows quick recovery of the routing performance between the source and destination. This requires us to provide online learning capabilities through intelligent sensing and computational mechanisms, and build routing protocols that can achieve security and resilience to adversarial attacks.

The secure routing against unknown jamming attacks in nature is a highly dynamic problem in distributed cognitive radio networks [4, 6, 9, 18]. We can categorize the current physical layer attacks into two types in distributed CR networks: jamming attacks and attacks for disturbing spectrum sensing.

In jamming, the attacker maliciously sends out packets to hinder legitimate participants in a communication session, resulting in a denial-of-service situation. The jammer sends continuously junk packets of data to a honest SU. The jammer can also disrupt communications by blasting a radio transmission, leading to the collision of packets in SUs [15].

Although simple energy-based detection and triangulation techniques can be used to detect jamming attacks, the time that it takes to pinpoint and ban the malicious user has a significant impact on the network performance. A jamming detection technique investigates the relationship between signal strength and packet delivery ratio [16]. If the signal strength is high, and yet the packet delivery ratio is low, then an honest SU can conclude that it is jammed unless one of its neighbors has a high signal strength and packet delivery ratio. Another technique called location consistency check is proposed to detect jamming where the location of SU neighbors is important [15]. Two strategies can be used to defend against jamming attacks. The first strategy is channel surfing or frequency hopping to escape jamming attack. The second strategy is spatial retreat where honest users change their location to escape the interference range imposed by the attacker. When no idle channel is available for honest SUs, the network topology is reconfigured to create new routing paths. In the case where an attacker sends packets on the same channel that a SU wants to use for transmission, the SU cannot pass the carrier-sensing and will be forced to back off in MAC (medium access control) layer, and consequently, the SU becomes aware of the jamming attack [15].

Primary user emulation (PUE) attack is a spectrum sensing attack that can be carried out by a malicious secondary user to emulate a PU or masquerade as a PU to obtain the resources of a given channel without having to share them with SUs [3]. As a result, the attacker can obtain full bands of a spectrum. An intelligent attacker jams the common control channel that is used to exchange sensing information between SUs in collaborative spectrum sensing. Furthermore, a malicious attacker can report false sensing data in the collaborative sensing environment [13]. These types of attacks can disturb the spectrum sensing function of honest SUs [15].

In [3], two approaches have been proposed to determine the location of the transmitter source of PUE attacks. One is distance ratio test which is based on received signal strength measurements and the other one is distance difference test which is based on signal phase difference. Both approaches follow a transmitter verification procedure. The procedure uses a location verification method to distinguish between primary signals and secondary signals masquerading as primary signals. In [13], a collaborative spectrum sensing technique has been proposed to defend against the malicious attackers who report false sensing data. However, the main drawback of this approach is the information overhead for exchanging collaborative information and the delay to recover routing paths.

In current literature, the effectiveness of these attacks in each layer and their defense methods are mostly studied independently [15]. Distributed routing faces various types of attacks that can lead to degraded network performance and reliability in distributed wireless networks such as ad hoc sensor networks [8]. Existing works in this area have focused on resource allocation techniques in multi-hop CR networks [5, 10]. In [7], private or public-key distribution schemes have been proposed to enhance the security in AODV routing. However, even with appropriate cryptographic techniques employed, the routing in CR networks is also vulnerable to attacks at the physical layer, which can critically deteriorate the performance and reliability. Smart attackers can coordinate attack activities in different layers to better achieve their goals and the capability of attackers. A cross-layer defense concept has been introduced to increase network efficiency through information exchange among different layers [7, 8, 12, 14]. Even though the mechanisms of multiple adversarial attacks have been well investigated in the literature, however, the way how to find secure routing paths against jamming attacks has not been much investigated in distributed CR networks. In this paper, we establish a dynamic routing game framework to provide a secure routing protocol to defend against malicious jamming attacks in distributed CR networks.

In the next section, we will describe the system model for dynamic secure routing.

4.3 System Model

In this section, we describe the network system model for the secure multi-hop cognitive radio networks. Let $\mathcal{G} := (\mathcal{N}, \mathcal{E})$ be a topology graph for a multi-hop cognitive radio network, where $\mathcal{N} = \{n_1, \ldots, n_N\}$ is a set of N secondary user nodes; and $\mathcal{E} := \{e_1, \ldots e_E\}$ is a set of E links connecting the secondary users. In addition, we let $\mathcal{M} := \{m_1, \ldots, m_K\}$ be a set of K primary users and $\mathcal{K}$ be the set of channels. We assume that a primary user is associated with one frequency channel, and the set of frequency channels is identical to the set of primary users. Primary user $m_p \in \mathcal{M}$ is associated with a channel, which can be in an either occupied state ($s_p = 1$) or an unoccupied state ($s_p = 0$). Let $\mathcal{S}_p := \{0, 1\}, m_p \in \mathcal{M}$, be the set of binary channel states of secondary user m_p. A system state $s = [s_p]_{m_p \in \mathcal{M}} \in \mathcal{S} := \prod_{p=1}^{K} \mathcal{S}_p$ is a collection of individual states of each primary user channel m_p. We assume that the system state s is an identically distributed random variable defined on the set $\mathcal{S}$.

In authors' earlier work [19], it is considered a secure routing problem in which secondary user $n_i \in \mathcal{N}$ needs to find an optimal routing path to its destination $n_i^d \in \mathcal{N}$. To initiate the routing process, secondary user n_i starts with an exploration of neighboring nodes for connections by sending them request messages. After n_i obtains a list of confirming nodes, it chooses a connecting node among them. Once the secondary user routes data to the selected node, the selected node initiates another exploration process to discover adjacent nodes along the direction towards the destination node. This process continues until the data of secondary user n_i reaches its destination. Denote by L_i the total number of explorations until a destination node is reached for secondary user n_i, and by $l_i \in \{0, 1, \ldots, L_i\}$ the l_ith exploration process with n_i as the initial node. Let $\mathcal{A}_{l_i} \subseteq \mathcal{N}$ be the set of nodes explored in l_ith exploration and $\mathcal{A} = \cup_{l_i=0}^{L_i} \mathcal{A}_{l_i} \subseteq \mathcal{N}$ be the total set of explored nodes along the path to the destination, including secondary user n_i and its destination node. By default, the exploration stage $l_i = 0$ refers to the initialization process of the routing, i.e., $\mathcal{A}_0$ is the singleton set containing secondary user n_i itself alone, and the L_ith exploration refers to the final stage of the routing, i.e., $\mathcal{A}_{L_i}$ is the singleton set which includes the destination node n_i^d. In this work, we consider single destination for each source node; however, it can be naturally extended to multiple destinations by defining a set of destination nodes.

Let $(n_i, l_i) \in \mathcal{A}_{l_i}$ be the node chosen after l_ith exploration to carry the data of node n_i. By default, at stage $l_i = 0$, we define $(n_i, 0) := n_i$, and at stage $l_i = L_i$, we define $(n_i, L_i) := n_i^d$. Let $\mathscr{P}_i(l_i, l_i') := \{(n_i, h), h = l_i, l_i + 1, \ldots, l_i'; l_i' > l_i, l_i, l_i' \in \{0, 1, \ldots, L_i\}\}$ be the set of connecting nodes along the multi-hop path between node (n_i, l_i) at stage l_i and node (n_i, l_i') at stage l_i' that routes the data of secondary user node n_i. Hence the path $\mathcal{Q}_i(l_i, l_i') \subset \mathscr{E}$ between (n_i, l_i) and (n_i, l_i') can be represented by the set of directed edges induced by $\mathscr{P}_i(l_i, l_i')$, i.e., $\mathcal{Q}_i(l_i, l_i') := \{\{(n_i, l_i), (n_i, l_i + 1)\}, \{(n_i, l_i + 1), (n_i, l_i + 2)\}, \ldots, \{(n_i, L_i - 1), (n_i, L_i)\}\}$. In particular, the complete multi-hop path from node n_i to its destination is denoted by $\mathcal{Q}_i(0, L_i)$. Note that the following composability property of $\mathscr{P}_i, \mathcal{Q}_i$, i.e., $\mathscr{P}_i(l_i, l_i') = \mathscr{P}_i(l_i, l_i'') \cup \mathscr{P}_{(n_i, l_i'')}(l_i'', l_i')$ and $\mathcal{Q}_i(l_i, l_i') = \mathcal{Q}_i(l_i, l_i'') \cup \mathcal{Q}_{(n_i, l_i'')}(l_i'', l_i')$, where $l_i < l_i'' < l_i'$, $l_i, l_i', l_i'' \in \{1, 2, \ldots, L_i\}$.

The multi-hop cognitive radio network can be subject to jamming attacks from multiple adversaries. Let $\mathscr{J} := \{1, 2, \ldots, J\}$ be the set of jammers in the CR network, and $\mathscr{R}_j, j \in \mathscr{J}$, be the set of nodes within the influence range of jammer j. Since the range of an attacker $\mathscr{R}_j \subset \mathcal{N}$ is wider than the effective jamming area of a power constrained jammer with directed antenna, without loss of generality, we can assume that the range $\mathscr{R}_j \subset \mathcal{N}$ of an attacker can only affect one local exploration stage, i.e., $\mathscr{R}_j \cup \mathcal{A}_h \neq \emptyset$ for some $h = 1, 2, \ldots, L_i - 1$, and $\mathscr{R}_j \cup \mathcal{A}_{h'} = \emptyset$, for all $h' \neq h, h' = 1, 2, \ldots, L_i - 1$. However, in the case where a jammer can cover a wide area enabled by multiple antenna, we can view the jammer equivalently as multiple local jammers over its covered area. Note that we have limited the stages between 1 and L_i because jamming at initial stage 0 can be easily detected by the source. The objective of a jammer j is to choose a node $r_j \in \mathscr{R}_j$ to jam and interrupt the data transmission of node $n_i \in \mathcal{N}$ to its destination. Let $\mathscr{J}_{l_i}$ be the set of

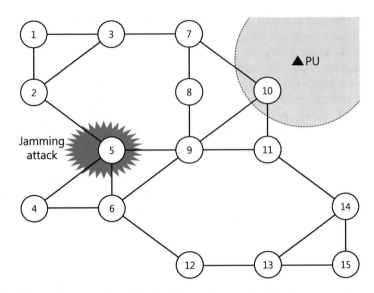

Fig. 4.1 The random network topology for 15 secondary nodes with a jammer, a primary user, a source, and a destination in $1 \times 1\,\mathrm{km}^2$ area

jammers whose influence range affects $\mathscr{A}_{l_i}$, i.e., $\mathscr{J}_{l_i} := \{j \in \mathscr{J}, \mathscr{R}_j \cap \mathscr{A}_{l_i} \neq \emptyset\}$. The action profile of jammers $\mathscr{J}_{l_i}$ is denoted by $\mathbf{r}_{l_i} := [r_j]_{j \in \mathscr{J}_{l_i}}$. The joint action profile of all the jammers is denoted by $\mathbf{r} = [r_j]_{j \in \mathscr{J}}$. The set of nodes within the jamming range by the set of jammers $\mathscr{J}_{l_i}$ is denoted by $\mathscr{R}_{l_i} := \cup_{j \in \mathscr{J}_{l_i}} \mathscr{R}_j$. Likewise, the set of nodes jammed by the set of jammers $\mathscr{J}$ is denoted by $\mathscr{R} = \cup_{j \in \mathscr{J}} \mathscr{R}_j$. In general, $\cup_{l_i=1}^{L_i} J_{l_i} \subseteq \mathscr{J}$; however, without loss of generality, we can assume that $\cup_{l_i=1}^{L_i} J_{l_i} = \mathscr{J}$ because we can exclude jammers that do not affect the routing of secondary user n_i.

The goal of secondary user node n_i is to choose an optimal path from the source to the destination that circumvents multiple jammers distributed along the path and yields the best routing performance as shown in Fig. 4.1.

In the next section, we derive the saddle-point in mixed strategies for proposed dynamic secure routing game.

4.4 Dynamic Game-Theoretic Model

The exploration and decision processes in the routing are composed of multi-stages. The strategic behaviors of SUs and jammers can be modeled using dynamic games. In this section, we define the utility functions and describe the game-theoretic model for secure routing in distributed cognitive radio networks with presence of multiple jamming adversaries.

4.4.1 Utility Function

As shown in [19], the performance of routing can be characterized by signal to interference plus noise ratio (SINR) that indicates physical-layer channel conditions, and routing delay that indicates the network-level congestions. Hence the utility function needs to capture these from the source to destination along the routing path. The average SINR from node $(n_i, l_i - 1)$ to node (n_i, l_i) is defined by

$$\gamma_{(n_i,l_i-1)}^{(n_i,l_i)} = \frac{\alpha_{(n_i,l_i-1)}^{(n_i,l_i)} \cdot P_{(n_i,l_i-1)}}{\sigma_{(n_i,l_i)}^2 + I_{(n_i,l_i-1)}^{(n_i,l_i)}}, \quad l_i = 1, 2, \ldots, L_i, \tag{4.1}$$

where $P_{(n_i,l_i-1)}$ is the transmit power of node $(n_i, l_i - 1)$ and $\alpha_{(n_i,l_i-1)}^{(n_i,l_i)} = \delta \cdot (d_{(n_i,l_i-1)}^{(n_i,l_i)})^{-\omega}$ is the channel gain between the node $(n_i, l_i - 1)$ and its selected node at (n_i, l_i) with $d_{(n_i,l_i-1)}^{(n_i,l_i)}$ the distance between $(n_i, l_i - 1)$ and (n_i, l_i); $\omega \in \mathbb{R}_{++}$ is the path loss exponent and $\delta \in R_{++}$ is the path loss constant; $\sigma_{(n_i,l_i)}^2$ is the variance of Gaussian noise at (n_i, l_i). The term $I_{(n_i,l_i-1)}^{(n_i,l_i)}$ is the interference perceived by neighboring nodes or malicious jammers at (n_i, l_i) and is given by

$$I_{(n_i,l_i-1)}^{(n_i,l_i)} = \sum_{n' \in \mathcal{N}_{(n_i,l_i)} \cup \mathcal{J}_{l_i} - \{(n_i,l_i-1)\}} \alpha_{n'}^{(n_i,l_i)} \cdot P_{n'}, \tag{4.2}$$

where $\mathcal{N}_{(n_i,l_i)} \subseteq \mathcal{N}$ is the set of nodes n' communicating with (n_i, l_i), including $(n_i, l_i - 1)$, with each using transmission power $P_{n'}$; and $\alpha_{n'}^{(n_i,l_i)}$ is the channel gain between n' and (n_i, l_i). The term $I_{(n_i,l_i-1)}^{(n_i,l_i)}$ represents the interference from the other nodes including jammers at (n_i, l_i). An adversary $j \, \mathcal{J}_{l_i}$ is assumed to jam one node at his maximum power $P_j^{\max}$, and depending on which node the jammer chooses, the term P_j in (4.2) takes the following form:

$$P_j = \begin{cases} P_j^{\max} & \text{if } r_j = (n_i, l_i) \\ 0 & \text{otherwise} \end{cases}, \tag{4.3}$$

for every $j \in \mathcal{J}_{l_i}$.

The primary goal of a jammer is to maximize the probability of packet error rate (PER) of SU n_i. The physical layer data communication PER $q_{(n_i,l_i-1)}^{(n_i,l_i)}$ between $(n_i, l_i - 1)$ and (n_i, l_i) is related to SINR $\gamma_{(n_i,l_i-1)}^{(n_i,l_i)}$ and the modulation and coding schemes as follows:

$$q_{(n_i,l_i-1)}^{(n_i,l_i)} = \phi_{(n_i,l_i-1)}^{(n_i,l_i)} \exp\left(-\varepsilon_{(n_i,l_i-1)}^{(n_i,l_i)} \cdot \gamma_{(n_i,l_i-1)}^{(n_i,l_i)}\right), \tag{4.4}$$

where $\phi_{(n_i,l_i-1)}^{(n_i,l_i)}$ is the maximum PER and $\varepsilon_{(n_i,l_i-1)}^{(n_i,l_i)}$ is a parameter depending on the modulation and coding schemes [11].

Another metric for routing is the communication delay at higher layers such as network or session layer. We recall the Pollaczek-Khinchin formula [2] for the M/G/1 queueing system and define the expected total packet delay $\tau^{(n_i,l_i)}_{(n_i,l_i-1)}$ perceived at node (n_i, l_i) as follows:

$$\tau^{(n_i,l_i)}_{(n_i,l_i-1)} = \frac{\eta_{(n_i,l_i-1)}\overline{X^2}_{(n_i,l_i)}}{2(1 - \rho^{(n_i,l_i)}_{(n_i,l_i-1)})} + \overline{X}_{(n_i,l_i)}, \tag{4.5}$$

where $\eta_{(n_i,l_i-1)}$ is the arrival rate of packets at the chosen node (n_i, l_i);

$$\rho^{(n_i,l_i)}_{(n_i,l_i-1)} := \eta_{(n_i,l_i-1)}/\mu_{(n_i,l_i)} = \eta_{(n_i,l_i-1)}\overline{X}_{(n_i,l_i)},$$

with $\mu_{(n_i,l_i)}$ being the service time at the node (n_i, l_i). $\overline{X}_{(n_i,l_i)}$ is the mean service time per packet at the chosen node and $\overline{X^2}_{(n_i,l_i)}$ is the expected variance of $\overline{X}_{(n_i,l_i)}$. When node (n_i, l_i) experiences a higher volume of incoming data, more delay will be perceived by $(n_i, l_i - 1)$ if (n_i, l_i) is selected.

Remark 1 Note that in (4.4) and (4.5), we have suppressed the dependence of the system state s and attacker's actions $\mathbf{r}_{l_i}$ in the notation. The parameters can be dependent on the primary user state s, in particular, the interference term and the path loss. In addition, the attackers' decisions $\mathbf{r}_{l_i}$ on which nodes to jam will affect the utility through (4.3) in (4.2).

The goal of SU n_i is to maximize its expected total utility U_i from the source to the destination, which is measured by the overall probability of successful transmission q and the total delay τ. Using (4.4) and (4.5), q and τ are defined, respectively, by $q(s, \mathscr{P}_i(0, L_i), \mathbf{r}) = \prod_{l_i=1}^{L_i} q^{(n_i,l_i)}_{(n_i,l_i-1)}$, and $\tau(s, \mathscr{P}_i(0, L_i), \mathbf{r}) = \sum_{l_i=1}^{L_i} \tau^{(n_i,l_i)}_{(n_i,l_i-1)}$. Note that minimizing (the expected value of) q is equivalent to minimizing (the expected value of)

$$\tilde{q}(s, \mathscr{P}_i(0, L_i), \mathbf{r}) := \sum_{l_i=1}^{L_i} \ln q^{(n_i,l_i)}_{(n_i,l_i-1)}.$$

Let $\lambda \in \mathbb{R}_{++}$ be a weighting parameter. We can use $\tilde{q}$ and τ to construct the multi-objective total utility as

$$U_i(s, \mathscr{P}_i(0, L_i), \mathbf{r}) = -\sum_{l_i=1}^{L_i} \left(\ln q^{(n_i,l_i)}_{(n_i,l_i-1)} + \lambda \tau^{(n_i,l_i)}_{(n_i,l_i-1)} \right),$$

$$= \sum_{l_i=1}^{L_i-1} u_{(n_i,l_i-1)}(s, (n_i, l_i), \mathbf{r}_{l_i}), \tag{4.6}$$

from which we can define the stage utility of node $(n_i, l_i - 1)$ selecting node (n_i, l_i) as follows: for $l_i = 1, \ldots, L_i$,

$$u_{(n_i, l_i - 1)}(s, (n_i, l_i), \mathbf{r}_{l_i}) := -\ln q_{(n_i, l_i - 1)}^{(n_i, l_i)} - \lambda \tau_{(n_i, l_i - 1)}^{(n_i, l_i)}. \tag{4.7}$$

Note that in general, λ can be interpreted as the Lagrange multiplier if we formulate the problem as optimizing $\tilde{q}$ subject to hard delay constraint. In addition, λ can be chosen to depend on the system state s. The stage utility and total utility are random variables as they are functions of s.

4.4.2 Maximin Problem

With (4.6) and (4.7), SU node n_i aims to find an optimal routing path that maximizes the expected total utility U_i along the route to the destination node, i.e.,

$$\max_{\mathcal{P}_i(0, L_i)} \mathbb{E}_s U_i(s, \mathcal{P}_i(0, L_i), \mathbf{r}). \tag{4.8}$$

The expected utility averages over the state space $\mathcal{S}$. It can also be interpreted as the time-average utility as s is an i.i.d. random variable over its support $\mathcal{S}$. However, the jammers intend to minimize the expected total utility (4.8) by choosing to attack the nodes between n_i and its destination. A jammer j at stage l_i can only jam a SU node $(n_i, l_i) \in \mathcal{A}_{l_i}$ within its range of influence $\mathcal{R}_j$. Hence, in our framework we will only need to concern with a set of SU nodes in $\mathcal{A}_{l_i}$ that have overlap with $\mathcal{R}_j$. It is easy to show that, for an adversary, an action in $\mathcal{R}_j \cup \mathcal{A}_{l_i}$ dominates an action in $\mathcal{R}_j \setminus \mathcal{A}_{l_i}$. Hence, without loss of generality, we can take $\mathcal{A}_{l_i}$ as the action spaces for attackers and the SU node at stage l_i.

The security strategy[1] for node n_i is a sequence of nodes $\{(n_i, l_i), l_i = 1, \ldots, L_i - 1\}$ that achieves the lower value

$$\underline{U}_i = \max_{\mathcal{P}_i(0, L_i)} \min_{\mathbf{r}} \mathbb{E}_s U_i(s, \mathcal{P}_i(0, L_i), \mathbf{r}). \tag{4.9}$$

The maximizing and minimizing strategies in (4.9) are denoted by $\mathcal{P}_i^*(0, L_i)$ and $\mathbf{r}^*$, respectively.

The problem (4.9) has a very special structure. Every SU node can only choose a connecting node for the next hop. The action space of SU n_i is limited to the set of neighboring nodes $\mathcal{A}_1$. In order for SU n_i to optimize over the entire path, it can select the best neighboring node that will yield the best utility in the future, or

---

[1]Note that "security strategy" is a game-theoretic term, referring the worst-case optimal strategies. It should not be interpreted as "cyber security strategy" as the term in computer science and engineering.

utility-to-go. This can be observed by decomposing (4.9) into two components: one is current stage utility and the other one is the utility-to-go. Given maximin strategy pairs $\mathscr{P}^*_{(n_i,1)}(1, L_i)$ and $\{\mathbf{r}^*_{l_i}\}^{L_i}_{l_i=1}$ for the exploration starting from SU node $(n_i, 1)$ to the destination, the problem (4.9) can be decomposed into

$$\underline{U}_i(\mathscr{P}^*_i(0, L_i), \mathbf{r}^*) = \max_{(n_i,1)} \min_{\mathbf{r}_1} \{\mathbb{E}_s u_i(s, (n_i, 1), \mathbf{r}_1)$$

$$+ \mathbb{E}_s \underline{U}_{(n_i,1)}(s, \mathscr{P}^*_{(n_i,1)}(1, L_i), \{\mathbf{r}^*_{l_i}\}^{L_i}_{l_i=1})\} \quad (4.10)$$

where the first term is the current utility and the second term is the utility-to-go. Solution to (4.10) yields a pair of maximin strategies $\mathscr{P}^*_i(0, 1)$ and $\mathbf{r}^*_1$ for node n_i. Hence the optimal path can be obtained as $\mathscr{P}^*_i(0, L_i) = \mathscr{P}^*_i(0, 1) \cup \mathscr{P}^*_{(n_i,1)}(1, L_i)$. Another special structure of the problem (4.9) is that, at each stage l_i, SUs face a distinct set of attackers. Hence, given a routing path, minimizing over $\mathbf{r}$ is the same as minimize over $\mathbf{r}_{l_i}$ at every stage. However, an intelligent attacker will minimize the current utility together with utility-to-go as in (4.10), taking into account the future routes of the SUs.

Leveraging these special properties of the problem, we arrive at a backward induction method to compute the maximin security strategies, which is summarized as follows:

Theorem 1 *Let $(\mathscr{P}^*_i(0, L_i), \mathbf{r}^*)$ be the pair of maximin solution to the problem (4.9) and $\underline{U}_{(n_i,l_i)}$ be the lower value of the utility from stage l_i to the destination, achieved under $\left(\mathscr{P}^*_{(n_i,l_i)}(l_i, L_i), \{\mathbf{r}^*_h\}^{L_i}_{h=l_i}\right)$ of this strategy pair. Then, the solution satisfies the following properties:*

$$\underline{U}_{(n_i,l_i)} = \max_{(n_i,l_i+1)} \min_{\mathbf{r}_{l_i+1}} \{\mathbb{E}_s u_{(n_i,l_i)}(s, (n_i, l_i + 1), \mathbf{r}_{l_i+1})$$

$$+ \mathbb{E}_s \underline{U}_{(n_i,l_i+1)}(s, \mathscr{P}^*_{(n_i,l_i+1)}(1, L_i), \{\mathbf{r}^*_h\}^{L_i}_{h=l_i+1})\} \quad (4.11)$$

$$l_i = 0, 1, 2, \ldots, L_i - 2.$$

$$\underline{U}_{(n_i,L_i-1)} = \mathbb{E}_s u_{(n_i,L_i-1)}(s, n^d_i, \mathbf{r}^*_{L_i}), \quad (4.12)$$

where $\mathbf{r}^*_{L_i} = \{n^d_i, \forall j \in \mathscr{J}_{L_i}\}$, and $\underline{U}_{(n_i,0)}(\mathscr{P}^*_i(0, L_i), \mathbf{r}^*) = \underline{U}_i$ defined in (4.9).

Proof At the penultimate stage $L_i - 1$, the chosen SUs $(n_i, L_i - 1) \in \mathscr{A}_{L_i-1}$ only need to connect to the destination n^d_i, which incurs a utility $\mathbb{E}_s u_{(n_i,L_i-1)}(s, n^d_i, \mathbf{r}^*_{L_i})$ where all adversary at the stage will jam the destination node. Using the argument (4.10), at stage $L_i - 2$, the utility function is composed of the stage utility and utility-to-go, which is given by (4.22). Using backward induction, we arrive at the relation (4.14) for an arbitrary stage l_i. $\square$

Remark 2 The above result provides a computation method to find the SU security strategies. The solution can be found by starting from the very last stage and propagates backwards to the initial stage. Such dynamic programming-like solutions have the properties of strong consistency or sub-game perfectness that will ensure the robustness of the solution.

In the next section, we will analyze zero-sum game and saddle-point in mixed strategies for dynamic secure routing game.

4.5 Zero-Sum Game and Saddle-Point in Mixed Strategies for Dynamic Secure Routing Game

The maximin problem in (4.9) yields the security strategies for the secondary users. On the other hand, intelligent jammers also intend to find their security strategies by solving an associated minimax problem as follows:

$$\overline{U}_i(\mathscr{P}_i^\circ(0, L_i), \mathbf{r}^\circ) = \min_{\mathbf{r}} \max_{\mathscr{P}_i(0, L_i)} \mathbb{E}_s U_i(s, \mathscr{P}_i(0, L_i), \mathbf{r}), \tag{4.13}$$

where the strategy pair $(\mathscr{P}_i^\circ(0, L_i), \mathbf{r}^\circ)$ is minimax solution to (4.13) and $\mathbf{r}^\circ$ is a security strategy for the jammers.

Theorem 2 *Let* $(\mathscr{P}_i^\circ(0, L_i), \mathbf{r}^\circ)$ *be the pair of minimax solution to the problem (4.13) and* $\overline{U}_{(n_i, l_i)}$ *be the upper value of the utility from stage* l_i *to the destination, achieved under* $\left(\mathscr{P}_{(n_i, l_i)}^\circ(l_i, L_i), \{\mathbf{r}_h^\circ\}_{h=l_i}^{L_i}\right)$ *of this strategy pair. Then, the solution satisfies the following properties:*

$$\overline{U}_{(n_i, l_i)} = \min_{\mathbf{r}_{l_i+1}} \max_{(n_i, l_i+1)} \left\{ \mathbb{E}_s u_{(n_i, l_i)}(s, (n_i, l_i + 1), \mathbf{r}_{l_i+1}) \right.$$

$$\left. + \mathbb{E}_s \overline{U}_{(n_i, l_i+1)} \left(s, \mathscr{P}_{(n_i, l_i+1)}^\circ(1, L_i), \{\mathbf{r}_h^\circ\}_{h=l_i+1}^{L_i} \right) \right\} \tag{4.14}$$

$$l_i = 0, 1, 2, \ldots, L_i - 2.$$

$$\overline{U}_{(n_i, L_i-1)} = \mathbb{E}_s u_{(n_i, L_i-1)}(s, n_i^d, \mathbf{r}_{L_i}^\circ), \tag{4.15}$$

where $\mathbf{r}_{L_i}^\circ = \{n_i^d, \forall j \in \mathscr{J}_{L_i}\}$, *and* $\overline{U}_{(n_i, 0)}(\mathscr{P}_i^\circ(0, L_i), \mathbf{r}^\circ) = \overline{U}_i$ *defined in (4.9).*

Following the problems defined in (4.9) and (4.13), we can define a zero-sum secure routing game between jamming adversaries and SUs for a SU source n_i, which is denoted by $\varXi_i := \{(n_i, \mathscr{N}, \mathscr{J}), (\mathscr{P}_i, \mathbf{r}), \mathscr{U}_i\}$. The solution to the zero-sum game $\varXi_i$ can be characterized by saddle-point equilibrium.

Definition 1 (Saddle-Point Equilibrium) Let $(\mathscr{P}_i^\star(0, L_i), \mathbf{r}^\star)$ be a feasible strategy pair. The zero-sum game $\varXi_i$ has a saddle-point in pure strategies, if the following inequalities hold, i.e.,

$$\mathbb{E}_s U_i(s, \mathscr{P}_i(0, L_i), \mathbf{r}^\star) \leq \mathbb{E}_s U_i(s, \mathscr{P}_i^\star(0, L_i), \mathbf{r}^\star) \leq \mathbb{E}_s U_i(s, \mathscr{P}_i^\star(0, L_i), \mathbf{r}),$$

for all feasible paths $\mathscr{P}_i(0, L_i)$ and jamming actions $\mathbf{r}$. The value of the game is given by

$$U_i^{\star} := \mathbb{E}_s U_i(s, \mathscr{P}_i^{\star}(0, L_i), \mathbf{r}^{\star}).$$

Saddle-point equilibrium, if exists, has many properties, which are described by the following theorem.

Theorem 3 *Suppose that the routing game Ξ_i for secondary user n_i has its upper value equal to its lower value, i.e., $\overline{U}_i = \underline{U}_i$ defined in (4.9) and (4.13), respectively. Then,*

   (i) the game has a saddle-point in pure strategies;
   (ii) an ordered pair of strategies provides a saddle point of Ξ_i if, and only if, the first of these is a security strategy for node n_i and the second one is a security strategies for the jammers;
(iii) $U_i^{\star}$ is uniquely given by $U_i^{\star} = \overline{U}_i = \underline{U}_i$.

The value of the game Ξ_i may not exist in pure strategies if $\overline{U}_i \neq \underline{U}_i$. In this subsection, we consider mixed strategies for the game Ξ_i. Let $\mathbf{f}_{i,l_i} = [f_{i,l_i}(n_{i'})]_{n_{i'} \in \mathscr{A}_{l_i}} \in \mathscr{F}_{i,l_i}$ be the mixed strategy at exploration l_i, which is a distribution over the action set $\mathscr{A}_{l_i}$, where

$$\mathscr{F}_{i,l_i} := \left\{ \mathbf{f}_{i,l_i} : \sum_{n_{i'} \in \mathscr{A}_{l_i}} f_{i,l_i}(n_{i'}) = 1, f_{i,l_i}(n_{i'}) \geq 0, \forall n_{i'} \in \mathscr{A}_{l_i} \right\}.$$

Likewise, let $\mathbf{g}_j = [g_j(r_j)]_{r_j \in \mathscr{R}_j} \in \mathscr{G}_j$ be the mixed strategies of the jammer, where

$$\mathscr{G}_j := \left\{ \mathbf{g}_j : \sum_{r_j \in \mathscr{R}_j} g_j(r_j) = 1, g_j(r_j) \geq 0, \ \forall r_j \in \mathscr{R}_j \right\}.$$

Let $\mathbf{F}_i = \{\mathbf{f}_{i,l_i}\}_{l_i=1}^{L_i} \in \mathscr{F}_i := \prod_{l_i=1}^{L_i} \mathscr{F}_{i,l_i}$ be mixed strategies for SUs in the secure routing game. Such strategies are also known as *behavioral strategies*, and $\mathscr{F}_i$ is the set of all feasible behavioral strategies. Note that, at stage L_i, the destination node n_i^d will be chosen. Hence the mixed strategy $\mathbf{f}_{i,L_i}$ at stage L_i is degenerated into a point distribution over the singleton action space $\mathscr{A}_{L_i}$. Let $\mathbf{G}_{l_i} = [\mathbf{g}_j : j \in \mathscr{J}_{l_i}] \in \prod_{j \in \mathscr{J}_{l_i}} \mathscr{G}_j$ and $\mathbf{G} = [\mathbf{g}_j]_{j \in \mathscr{J}} \in \mathscr{G} := \prod_{j \in \mathscr{J}} \mathscr{G}_j$. The average stage utility for node i to send data from $(n_i, l_i - 1)$ to (n_i, l_i) in mixed strategies is

$$\mathfrak{u}_{(n_i, l_i)}(s, \mathbf{f}_{i,l_i+1}, \mathbf{G}_{l_i+1}) = \mathbb{E}_{s, \mathbf{f}_{i,l_i}, \mathbf{G}_{l_i}} u_{(n_i, l_i)}(s, (n_i, l_i + 1), \mathbf{r}_{l_i+1}). \quad (4.16)$$

The total average utility for node i to reach its destination node in mixed strategies is

$$\mathbb{U}_i(s, \mathbf{F}_i, \mathbf{G}) = \sum_{l_i=1}^{L_i-1} \mathfrak{u}_{(n_i, l_i-1)}(s, \mathbf{f}_{i,l_i}, \mathbf{G}_{l_i}). \quad (4.17)$$

The upper value and lower value of the game are given by

$$\overline{\mathbb{U}}_i := \max_{\mathbf{F}_i \in \mathscr{F}_i} \min_{\mathbf{G} \in \mathscr{G}} \mathbb{E}_s \mathbb{U}_i(s, \mathbf{F}_i, \mathbf{G}),$$

$$\underline{\mathbb{U}}_i := \min_{\mathbf{F}_i \in \mathscr{F}_i} \max_{\mathbf{G} \in \mathscr{G}} \mathbb{E}_s \mathbb{U}_i(s, \mathbf{F}_i, \mathbf{G}),$$

and the corresponding maximizing and minimizing pairs are $(\mathbf{F}_i^*, \mathbf{G}^*)$ and $(\mathbf{F}_i^\circ, \mathbf{G}^\circ)$, respectively.

Theorem 4 *Consider the game $\varXi_i$ described in this section. Then,*

(i) *$\varXi_i$ has a saddle-point in mixed strategies $(\mathbf{F}_i^*, \mathbf{G}^*)$, satisfying*

$$\mathbb{E}_s \mathbb{U}_i(s, \mathbf{F}_i, \mathbf{G}^*) \le \mathbb{E}_s \mathbb{U}_i(s, \mathbf{F}_i^*, \mathbf{G}^*) \le \mathbb{E}_s \mathbb{U}_i(s, \mathbf{F}_i^*, \mathbf{G}),$$

for all feasible mixed strategies $\mathbf{F}_i \in \mathscr{F}_i, \mathbf{G} \in \mathscr{G}$.

(ii) *The zero-sum game has a value in mixed strategies uniquely given by $\mathbb{U}_i^* = \overline{\mathbb{U}}_i = \underline{\mathbb{U}}_i$.*

(iii) *A pair of mixed strategies provides a saddle point for $\varXi_i$ if, and only if, the first of these is a mixed security strategy for node n_i, and the second one is the mixed security strategy for the jammers.*

Proof The multi-hop game $\varXi_i$ with finite number of players and finite discrete pure-strategy action spaces. Using Theorem 2.4 in [1], there exists a saddle-point in mixed strategies. In addition, due to the multi-stage structure of the game, the behavioral strategies can be found using backward induction. The results in (ii) and (iii) then directly follow from Corollary 2.3 in [1]. □

Theorem 4 provides the existence of saddle-point equilibrium, and the existence and uniqueness of the value of the game $\varXi_i$. The saddle-point in mixed strategies can be computed using backward induction similar to the security strategies. Let val$(\cdot)$ be the value operator and the value of the game $\mathbb{U}_i^*$ can be written as

$$\mathbb{U}_i^* = \text{val}\{\mathbb{E}_s \mathbb{U}_i(s, \mathbf{F}_i, \mathbf{G})\}, \tag{4.18}$$

$$(\mathbf{F}_i^*, \mathbf{G}^*) \in \arg \text{val}\{\mathbb{E}_s \mathbb{U}_i(s, \mathbf{F}_i, \mathbf{G})\}, \tag{4.19}$$

where $(\mathbf{F}_i^*, \mathbf{G}^*) \in \mathscr{F}_i \times \mathscr{G}$ is the saddle-point equilibrium strategy pair achieved under the val operator. We can use the result described in Theorem 5 to compute the strategies.

Theorem 5 *Let $U_{(n_i,l_i)}^*$ be the value of the game $\varXi_{(n_i,l_i)}$, which is a truncated game of $\varXi_i$, which starts from stage l_i to the destination. Then, the saddle-point equilibrium satisfies*

$$U_{(n_i,l_i)}^* = \text{val}\left\{ \mathbb{E}_s u_{(n_i,l_i)}\left(s, \mathbf{f}_{l_i+1}, \{\mathbf{g}_j\}_{j \in \mathscr{J}_{l_i+1}}\right) \right.$$

$$\left. + \mathbb{E}_s U_{(n_i,l_i+1)}^*\left(s, \mathbf{F}_{(n_i,l_i+2)}^*, \{\mathbf{g}_j^*, j \in \mathscr{J}_h\}_{h=l_i+2}^{L_i}\right) \right\}$$

$$\tag{4.20}$$

$$(\mathbf{f}_{l_i+1}^*, \{\mathbf{g}_j^*, j \in \mathscr{J}_{l_i+1}\}) \in arg\ val \left\{ \mathbb{E}_s \mathbb{u}_{(n_i, l_i)} \left(s, \mathbf{f}_{l_i+1}, \{\mathbf{g}_j\}_{j \in \mathscr{J}_{l_i+1}} \right) \right.$$

$$\left. + \mathbb{E}_s \mathbb{U}_{(n_i, l_i+1)}^* \left(s, \mathbf{F}_{(n_i, l_i+1)}^*, \{\mathbf{g}_j^*, j \in \mathscr{J}_h\}_{h=l_i+1}^{L_i} \right) \right\}$$

$$(4.21)$$

$$l_i = 0, 1, 2, \ldots, L_i - 2.$$

$$U_{(n_i, L_i-1)}^* = \mathbb{E}_s \mathbb{u}_{(n_i, L_i-1)}(s, \mathbf{f}_{L_i}^*, \{\mathbf{g}_j^*, j \in \mathscr{J}_{L_i}\}), \qquad (4.22)$$

where the saddle-point strategies at last exploration stage L_i are given by

$$\mathbf{f}_{L_i}^* = n_i^d, \quad w.\ p.\ 1,$$

$$\mathbf{g}_j^* = n_i^d, \quad w.\ p.\ 1, \quad \forall\ j \in \mathscr{J}_{L_i},$$

The saddle-point equilibrium computed from (4.21) forms a behavioral mixed strategy equilibrium $(\mathbf{F}_{(n_i, l_i)}^*, \{\mathbf{g}_j^*, j \in \mathscr{J}_h\}_{h=l_i}^{L_i})$ associated with the truncated game. With $l_i = 0$, $\mathbb{U}_{(n_i, 0)}^* = \mathbb{U}_i^*$ defined in (4.18) and $(\mathbf{F}_{(n_i, 0)}^*, \{\mathbf{g}_j^*, j \in \mathscr{J}_h\}_{h=0}^{L_i})$ coincides with $(\mathbf{F}_i^*, \mathbf{G}^*)$ defined in (4.19).

Proof Based on Theorem 4, there exists a game value and mixed saddle-point strategies for each truncated game. At last stage, n_i^d is chosen for connection and jamming. Using backward induction, we obtain the results above. □

Novel distributed learning algorithms provide us tools towards design and implementation of practical protocols [19]. Depending on the initial knowledge of secondary users, we can adopt fictitious play for experienced nodes that have the knowledge of their utility functions and adversaries. On the other hand, secondary users without initial knowledge can use distributed Boltzmann-Gibbs learning algorithms due to its capability of estimating the expected utility and making decision without the complete knowledge of adversaries and the payoff structure. The Boltzmann-Gibbs learning starts with a local procedure defining the set of nodes for direct links, calculates his payoff, and selects the best routing node at the next hop with the maximum utility, i.e., the total path utility U_i, and dynamically updates whenever the SUs acquire the new estimate system knowledge. Once the source to the destination pair is determined, SUs learn and update their total path payoffs. Each node updates its mixed strategies by the Boltzmann-Gibbs learning until the process converges to an ϵ-saddle-point equilibrium. Under different learning rates on the strategies and the average payoffs, the learning algorithm converges at each exploration site. Here, we have assumed that the convergence of the learning algorithms is faster than the speed of exploration. The employment of such learning algorithms at each stage will lead to the mixed-strategy saddle-point equilibrium discussed.

Theorem 6 *Suppose that payoff learning in the Boltzmann-Gibbs learning algorithm is faster than the strategy learning at each stage for every SU (n_i, l_i), $l_i = 0, 1, \ldots, L_i$, and jammer $j \in \mathscr{J}_{l_i}$. Then, if the multi-stage learning algorithm converges, it yields a mixed-strategy saddle-point equilibrium of the game Ξ_i as $t \to \infty$ and $\epsilon \to 0$.*

Proof At the last stage L_i, the mixed-strategies for SU node $(n_i, L_i - 1)$ and adversary $j, j \in L_i$ are given in (4). The learning at stage $L_i - 1$ under the B-G algorithm will lead to $\hat{\mathbf{u}}_{i,l_i}^k \to \mathbb{U}_{(n_i,L_i-2)}^*$ as $k \to \infty$, and $\mathbf{f}_{i,l_i}^k \to \mathbf{f}_{l_i}^*$, $\mathbf{g}_{j,l_i}^k \to \mathbf{g}_j^*$, $j \in \mathscr{J}_{l_i}$ (see [20] and [17]). After the learning at stage $L_i - 1$ reaches its steady state, we can conclude the same convergence results for stage $L_i - 2$. Hence, by backward induction, at initial stage, the source n_i will learn the game value $\mathbb{U}_i^*$ and saddle-point equilibrium $(\mathbf{F}_i, \mathbf{G})$ as t is sufficiently long and $\epsilon \to 0$ across the stages.   □

The proposed secure routing algorithm above is used to combat jamming attacks in distributed cognitive radio networks. The state of primary channels evolves in the network and hence the routing is state-dependent. The secure routing game can spatially circumvent jammers along the routing path and can learn to defend against malicious attackers as the state changes. In the algorithm, the first step of the secure routing game is to establish an initial routing path from the source node to the destination node by initializing local game for direct links and for global paths. Once the initial path is established and it is converged to the saddle-point equilibrium by the Boltzmann-Gibbs learning, the source node starts transmitting packet data through the initial path. If the packet data transmission is terminated, then the secure routing game is concluded for the session transmission and another game will start for transmitting other session.

4.6 Simulation Results

In this experiment, we used network simulator 2 (NS-2) to determine the performance of proposed Distributed Secure Routing Protocol (DSRP). In simulation set up, we used an omni-directional antenna with two-ray-ground as propagation model. IEEE 802.11b protocol is employed for medium access by wireless nodes, where we used standard values for all the parameters defined by the IEEE802.11b. Each wireless node interface has queue length of 50 packets and follows drop tail queuing mechanism. Nodes are placed according to the topology depicted in Fig. 4.1 in the area $1000 \times 1000\,\text{m}^2$. Each wireless node's transmission range is set to 250 m, while the wireless link data rate used is 1 Mbps. Transmission Control Protocol (TCP) is employed as transport layer protocol, while File Transfer Protocol (FTP) is used as a source for the generation of data traffic. Different pairs of source-destination are used with single data flow. We compared our proposed DSRP with the Ad-hoc On-demand Distance Vector (AODV) routing protocol. The proposed DSRP targets to reduce the path recovery in case of node failures to defense any jamming attacks. In the proposed DSRP, each node tends to employ learning from the environment. In this experiment, we tend to show the effectiveness of proposed DSRP against node failure such as jamming attack, Primary User Emulation (PUE) attack, and primary users.

In Fig. 4.2, instantaneous throughput performance of the proposed DSRP is compared with throughput performances of no-attack and AODV when a jamming attacks. The initial path was from source node 1 to destination node 15 via 1, 2, 5,

Fig. 4.2 Instantaneous throughput performance of the proposed DSRP algorithm, when a jammer attacks node 2 at time between 15 and 20, compared with no-attack and AODV. The source is node 1 and the destination is node 15 in Fig. 4.1

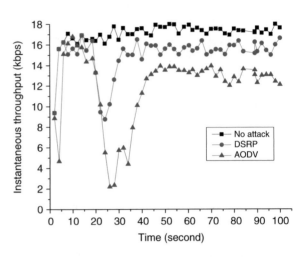

Fig. 4.3 Instantaneous throughput performance of the proposed DSRP algorithm against different jamming attacks such as a jammer attack and a PUE attack

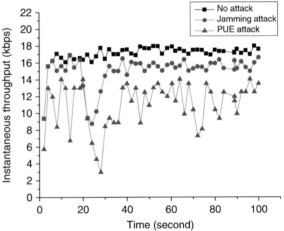

9, 11, 14, 25 in Fig. 4.1. The jamming attack is launched at time between 15 and 20 s, the attack compromised the node 2 and the path to destination via node 2 is no more valid. The alternate path that can lead to destination go via node 4, 7, 11, 12, 14. Instantaneous throughput degrades at the time attack is launched, the proposed DSRP fast recovers the path while the typical AODV is slower as shown in Fig. 4.2. Further, the instantaneous throughput of DSRP is still better than AODV, because DSRP employs continuous learning mechanism at each node and keeps updating the fluctuations of link conditions.

The impact of different attacks such as jamming attacks and PUE attacks affects throughput of the network and results in higher delay and lower throughput. Figure 4.3 depicts the comparison of instantaneous throughput performances between different attacks such as a jamming and a PUE attack and shows how much delay is encountered for path recovery and how much throughput degrades in the simulations. Source and destination nodes are 7 and 15, respectively, and the

Fig. 4.4 Instantaneous throughput performance of the proposed DSRP algorithm and AODV under multiple attacks

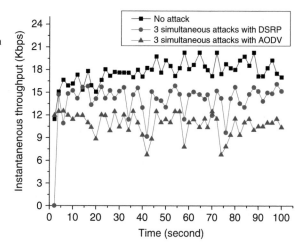

attack is launched in the time between 15 and 20 s. In Fig. 4.3, node 10 becomes un-available to connect due to PUE attack, so the route establishment delay is encountered. In the PUE attack, the path established is changed from original path, when there is no attack (10, 11, 14), to secondary path (4, 8, 11, 14). Since, the path established is still better, so the throughput is not degraded too much. While in case of jamming attack the path followed is longer and throughput is degraded more than previous case. In our topology, the jammer node (node 11) becomes unavailable due to attack. So, after delay of route establishment, data transmission starts, but the path followed is longer (4, 8, 6, 9, 12) as compared to previous case.

In Fig. 4.4, we demonstrate the comparison of the proposed DSRP and AODV under multiple jamming attacks. In this scenario, node 10 is in range of primary user and node 5 and node 11 are under simultaneous jamming attacks which decreases their utilities at 35 s, so the path to destination becomes (4, 8, 6, 9, and 12). The impact of multiple attacks degrades throughput as shown in Fig. 4.4, but the proposed DSRP still fast recover and outperforms AODV.

Figure 4.5 shows the end-to-end recovery delay which is the required time to establish route from source to destination after any node on the initial path is attacked by a jammer. Figure 4.5 compares the end-to-end recovery delays between the proposed DSRP and AODV. The increasing number of jammers affects the delay required for the alternate path recovery due to the route establishment overhead and change in routing paths. Proposed DSRP takes less time to recover compared to the typical AODV routing protocol.

Figure 4.6 compares the overhead required to recover the end-to-end path between the proposed DSRP and AODV. Since AODV broadcasts route error messages to all neighbors and the source node reinitiate the route discovery to recover a new path whenever there is a jammer, the overhead is steeply increase by increasing number of jammer while the proposed DSRP algorithm recover it locally and dynamically, the overhead for recover paths is smaller than the overhead of AODV.

Fig. 4.5 The end-to-end recovery delay time for the proposed DSRP compared with AODV versus the number of jammers

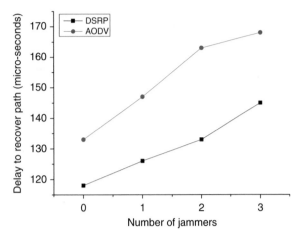

Fig. 4.6 The overhead for recovery paths for the proposed DSRP compared with AODV versus the number of jammers

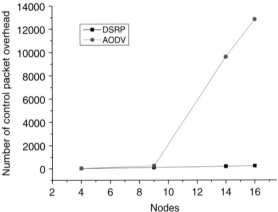

4.7 Conclusion

In this paper, we have presented an on-line secure routing scheme based on a dynamic zero-sum game framework. The proposed secure routing game model dynamically defenses against malicious attacks such as jamming in cognitive radio network in distributed manner. The proposed algorithm defenses against unknown attackers for their routing and minimizes the packet error probability and delay along the routing path from the source to the destination in optimal and distributed manner. Unlike typical distributed routing algorithm such as an AODV routing scheme, the suggested secure routing game algorithm supports a novel recovery of routing path failure against unknown attackers. We have seen that the proposed scheme allows nodes to recover from path failure caused by unknown attacks and that the delay and throughput performances are optimized in comparison to its

AODV shortest path routing counterpart. The future work includes a generalization of the secure routing framework for applications in 5G license-assisted access distributed networks and sensor networks.

References

1. Başar, T., Olsder, G.J.: Dynamic Noncooperative Game Theory. SIAM Series in Classics in Applied Mathematics. SIAM, Philadelphia (1999)
2. Chan, W.C., Lu, T.C., Chen, R.J.: Pollarczek-Khinchin formula for the M/G/1 queue in discrete time with vacations. In: IEE Proceedings of Computers and Digital Techniques, vol. 144, pp. 222–226 (1997)
3. Chen, R., Park, J.M., Reed, J.H.: Defense against primary user emulation attacks in cognitive radio networks. IEEE J. Sel. Areas Commun. **26**, 25–37 (2008)
4. Clancy, T.C., Goergen, N.: Security in cognitive radio networks: threats and mitigation. In: Proceedings of the 3rd International Conference on CrownCom08, Singapore (2008)
5. Ding, L., Melodia, T., Batalama, S., Medley, M.J.: ROSA: distributed joint routing and dynamic spectrum allocation in cognitive radio ad hoc networks. In: Proceedings of ACM MSWiM, Canary Islands, Spain (2009)
6. Hossain, E., Niyato, D., Han, Z.: Dynamic Spectrum Access in Cognitive Radio Networks. Cambridge University Press, Cambridge (2009)
7. Hu, Y.-C., Perrig, A.: A survey of secure wireless ad hoc routing. IEEE Secur. Priv. **2**, 28–39 (2004)
8. Karlof, C., Wagner, D.: Secure routing in wireless sensor networks: attacks and countermeasures. In: Proceedings of IEEE International Workshop on Sensor Network Protocols and Applications, Anchorage, AK, USA (2003)
9. Saad, W., Zhu, Q., Başar, T., Han, Z., Hjorungnes, A.: Hierarchical network formation games in the uplink of multi-hop wireless networks. In: Proceedings of IEEE Globecom, Honolulu, Hawaii (2009)
10. Shi, Y., Hou, Y.T.: A distributed optimization algorithm for multi-hop cognitive radio networks. In: Proceedings of International Conference on Computer Communications, Phoenix, AZ, USA (2008)
11. Tague, P., Slater, D., Noubir, G., Poovendran, R.: Linear programming models for jamming attacks on network traffic flows. In: Proceedings of WiOpt'08, Berlin, Germany (2008)
12. Tague, P., Nabar, P., Ritcey, J.A., Poovendram, R.: Jamming-aware traffic allocation for multiple-path routing using portfolio selection. IEEE/ACM Trans. Networking **19**, 184–194 (2011)
13. Wang, W., Li, H., Sun, Y., Han, Z.: Securing collaborative spectrum sensing against untrustworthy secondary users in cognitive radio networks. EURASIP J. Adv. Signal Process. **2010**, 106–117 (2010)
14. Wang, W., Sun, Y., Li, H., Han, Z.: Cross-layer attack and defense in cognitive radio networks. In: Proceedings of IEEE Globecom, Miami, Florida, USA (2010)
15. Wassim, E., Haidar, S., Mohsen, G.: Survey of security issues in cognitive radio networks. J. Internet Technol. **12**, 181–198 (2011)
16. Xu, W., Trappe, W., Zhang, Y., Wood, T.: The feasibility of launching and detecting jamming attacks in wireless networks. In: Proceedings of ACM MobiHoc, Urbana, IL, USA (2005)
17. Zhu, Q., Tembine, T., Başar, T.: Heterogeneous learning in zero-sum stochastic games with incomplete information. In: Proceedings of IEEE Conference on Decision and Control, Atlanta, Georgia (2010)
18. Zhu, Q., Yuan, Z., Song, J.B., Han, Z., Başar, T.: Dynamic interference minimization routing game for on-demand cognitive pilot channel. In: Proceedings of IEEE Globecom, Miami, Florida (2010)

19. Zhu, Q., Song, J.B., Başar, T.: Dynamic secure routing game in distributed cognitive radio networks. In: Proceedings of IEEE Globecom, Houston, Texas (2011)
20. Zhu, Q., Tembine, H., Başar, T.: Distributed strategic learning with application to network security. In: Proceedings of IEEE American Control Conference, San Francisco, CA (2011)
21. Zhu, Q., Yuan, Z., Song, J.B., Han, Z., Başar, T.: Interference aware routing game for cognitive radio multi-hop networks. IEEE J. Sel. Areas Commun. **30**, 2006–2015 (2012)

Chapter 5
Content Sponsoring with Inter-ISP Transit Cost

Abylay Satybaldy and Changhee Joo

5.1 Introduction

As demand for mobile data increases, Internet service providers (ISPs) are turning to new types of smart data pricing to bring in additional revenue and to expand the capacity of their current network [7]. One way to keep up funding such investment is content sponsorship. Content providers (CPs) split the cost of transferring mobile data traffic, and sponsor the user's access to the content by making direct payment to the ISPs. For example, GS Shop, a Korea TV home shopping company, has partnered with SK Telecom to sponsor data incurred from its application, so consumers are incentivized to continue browsing and making purchases from their mobile devices without ringing up data charges [1]. Content sponsoring may benefit all players in the market: the ISPs can generate more revenue with CP's subsidies, and users can enjoy free or low-cost access to certain services, which in turn increases the demand and attracts more traffic, resulting in higher revenue of the CP.

There are several studies on content sponsoring despite a short history. Most of the works either focus on a simple model with a single ISP and a single CP interacting in a game theoretic setting or consider Quality-of-Service (QoS) prioritization and its implications for net neutrality [4, 5]. In a two-sided market with a single ISP providing connection between CPs and EUs, profit maximization of the players under sponsoring mobile data has been studied in [2, 8]. In [2], single monopolistic ISP determines optimal price to charge the CPs and the EUs, while the authors in [8] study the contractual relationship between the CPs and the ISP under a similar model. Nevertheless, none of them consider the interaction between multiple ISPs. Although the authors in [9] propose a model with a transit

A. Satybaldy · C. Joo (✉)
Ulsan National Institute of Science and Technology (UNIST), Ulsan, South Korea
e-mail: abylay@unist.ac.kr; cjoo@unist.ac.kr

© Springer Nature Switzerland AG 2019
J. B. Song et al. (eds.), *Game Theory for Networking Applications*,
EAI/Springer Innovations in Communication and Computing,
https://doi.org/10.1007/978-3-319-93058-9_5

ISP and a user-facing ISP, their understanding of the interaction between these non-cooperative ISPs is limited to the environments without content sponsoring. Other works, e.g. [3, 10], have analyzed content sponsorship from the economic point of view. They examine the implications of sponsored data on the CPs and the EUs, and identify how sponsored data influence the CP inequality.

In many Internet markets, there are multiple ISPs that cooperate to provide end-to-end connectivity service between the CPs and the EUs, in which case the assumption of a single representative ISP no longer holds. Since each ISP aims to maximize its own profit, the establishment of interconnection among multiple ISPs is a thorough process that depends on specific profit sharing/inter-charging arrangements.

As the most commercial traffic originates from the CPs and terminates at the EUs, some ISPs positioned on the middle of the traffic delivery chain will have more power and request a transit-price. An ISP serving a large population of users might have a dominant influence in determining the transit price paid by other relatively weak ISPs for traffic delivery. For example, a large entertainment company Netflix directly uses the service provided by ISPs such as Level 3, which is connected with residential broadband ISPs like Comcast to get access to the customers [6]. Level 3 charges Netflix and Comcast charges the users. Netflix may partially or fully sponsor its traffic, which is likely to increase the amount of traffic through both ISPs. Due to high traffic volume, the access ISP (Comcast) may require additional transit price for traffic delivery, which will impact on the pricing decision at Level 3 and subsequently on the sponsoring decision at Netflix. In this work, we are interested in the dynamics between the players with focus on content sponsoring and transit pricing. To this end, we study the interplay among two ISPs, CP, and EU, where each player selfishly maximizes its own profit. We model this non-cooperative interaction between ISP_1, ISP_2, CP, and EU as a four-stage Stackelberg game. Specifically, in our model, we assume that the EU-facing ISP has a dominant power and can be considered as the game leader who decides the transit cost preceding the choice of the follower ISP. We aim to understand the behaviors of the players in non-cooperative equilibrium and their decisions to maximize their own utility.

The rest of the paper is organized as follows. We present the basic system model in Sect. 5.2, and investigate the strategies of the CP, the EU, and the ISPs to maximize their utility in Sect. 5.3. Numerical results are presented in Sect. 5.4, followed by the conclusion and future work in Sect. 5.5.

5.2 Two-ISP Pricing Model

We consider an Internet market model with one CP and two ISPs as shown in Fig. 5.1. Two interconnected ISPs have their own cost structures and each provides connectivity to either the CP or the EU. The CP-facing ISP (ISP_1) obtains its profits by directly charging the CP (CP) by p_{cp} for per unit traffic while the EU-facing ISP (ISP_2) charges the EU (EU) by p_{eu} for per unit traffic. Further ISP_2 charges ISP_1

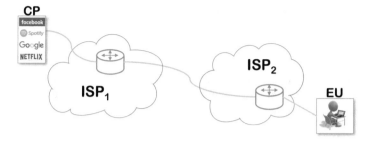

Fig. 5.1 Two-sided Internet market

with transit-price p_{tr} for traffic delivery. Let m_1 and m_2 denote the marginal costs of traffic delivery for ISP_1 and ISP_2, respectively. We denote x as the traffic amount of flow between CP and EU.

We assume that the players in this non-cooperative game make decisions in four stages as follows:

1. ISP_2 sets prices p_{eu} and p_{tr} to charge EU and ISP_1, respectively.
2. ISP_1 determines the optimal value of p_{cp} to charge CP.
3. CP decides how much content to sponsor, i.e., the value of s.
4. The traffic volume is decided by both EU and CP.

Each player selfishly maximizes its own profit subject to the others' decisions. We model this non-cooperative interaction as a four-stage Stackelberg game and use the backward induction method to find optimal strategy of each player.

Let us define the utility of EU by the multiplication of a scaling factor $\sigma_{eu} \geq 0$ and a utility-level function. The utility represents user's desire to obtain traffic. We assume a concave and non-decreasing function $u_{eu}(x)$ with decreasing marginal satisfaction, i.e., $u_{eu}(x) = \frac{x^{1-\alpha_{eu}}}{1-\alpha_{eu}}$ with parameter $\alpha_{eu} \in (0, 1)$. Given unit price p_{eu} that ISP_2 charges user, EU will maximize its utility minus the payment by solving

$$(EU-P) \quad \max_{x} \quad \sigma_{eu} \cdot u_{eu}(x) - (1-s) \cdot x \cdot p_{eu},$$

$$s.t. \quad x \geq 0, \tag{5.1}$$

where $s \in [0, 1]$ denotes the sponsored percentage, and $(1-s) \cdot x \cdot p_{eu}$ denotes the payment of EU to ISP_2. The solution x_{eu}^* to (5.1) can be obtained as $x_{eu}^*(s, p_{eu}) = \left(\frac{\sigma_{eu}}{(1-s)p_{eu}}\right)^{\frac{1}{\alpha_{eu}}}$.

Similarly, we model the behavior of CP. The utility of CP is given by $\sigma_{cp}u_{cp}(x)$, where $\sigma_{cp} \geq 0$ is a scaling factor (e.g., the popularity of the content) and $u_{cp}(x)$ is a concave utility-level function $u_{cp}(x) = \frac{x^{1-\alpha_{cp}}}{1-\alpha_{cp}}$ with parameter $\alpha_{cp} \in (0, 1)$. CP will maximize its payoff by solving

$$(CP-P) \quad \max_{x,s} \quad \sigma_{cp} \cdot u_{cp}(x) - s \cdot x \cdot p_{eu} - x \cdot p_{cp},$$

$$s.t. \quad x \geq 0 \quad and \quad 0 \leq s \leq 1. \tag{5.2}$$

In the objective, the first term denotes its utility, the second term denotes the cost due to sponsorship, and the third term is from the network usage cost to ISP_1. Given s, p_{cp}, and p_{eu}, it can be easily shown that the optimal amount of traffic for CP is

$$x_{cp}^*(s, p_{cp}, p_{eu}) = \left(\frac{\sigma_{cp}}{s p_{eu} + p_{cp}}\right)^{\frac{1}{\alpha_{cp}}}.$$

Since ISP_1 obtains its revenue from charging CP, it decides the optimal value of p_{cp} to maximize its total profit as

$$(ISP_1-P) \quad \max_{p_{cp}} \quad (p_{cp} + s^* \cdot p_{eu} - p_{tr} - m_1) \cdot x^*(p_{cp}, p_{eu}),$$

$$s.t. \quad p_{cp} \geq 0, \tag{5.3}$$

where m_1 is the marginal cost for traffic delivery and thus $p_{cp} + s^* \cdot p_{eu} - p_{tr} - m_1$ is the net-gain of ISP_1 per unit traffic.

ISP_2 obtains its revenue from charging ISP_1 with transit-price p_{tr} and charging EU with traffic-price p_{eu}. Therefore, in order to maximize its total profit, it will solve

$$(ISP_2-P) \quad \max_{p_{eu}, p_{tr}} \quad ((1 - s^*) \cdot p_{eu} + p_{tr} - m_2) \cdot x^*(p_{cp}, p_{eu}),$$

$$s.t. \quad p_{eu} \geq 0 \quad and \quad p_{tr} \geq 0, \tag{5.4}$$

where m_2 is the marginal cost for traffic delivery.

Through the sequential decision, we investigate the interactions of the players described in (5.1), (5.2), (5.3), (5.4), and find the optimal strategies for pricing and sponsoring.

5.3 Strategies for Utility Maximization

In this section, we sequentially find the optimal strategies of CP, ISP_1, and ISP_2 by exploiting the backward induction.

5.3.1 Sponsoring of Content Provider

Note that each solution to (5.1) and (5.2) results in user-side traffic demand x_{eu}^* and CP-side traffic amount x_{cp}^*, respectively, and the actual traffic amount x^* between

CP and EU will be determined by their minimum, i.e., $x^* = \min\{x^*_{cp}, x^*_{eu}\}$. In general $x^*_{eu} \neq x^*_{cp}$. For instance, a certain website may restrict the number of simultaneous on-line clients, which implies $x^*_{cp} \leq x^*_{eu}$.

Suppose that p_{eu} and p_{cp} are given. The actual traffic $x^*(s)$ will be determined by the sponsoring rate s, and CP will decide its optimal sponsored percentage s^* by solving the following problem:

$$(\boldsymbol{CP - P}) \quad \max_{s} \quad \sigma_{cp} \cdot u_{cp}(x^*(s)) - s \cdot x^*(s) \cdot p_{eu} - x^*(s) \cdot p_{cp},$$

$$s.t. \quad 0 \leq s \leq 1. \tag{5.5}$$

We assume $\alpha_{eu} = \alpha_{cp} = \alpha \in (0, 1)$, i.e., EU and CP utility components have the same utility shape. This assumption is reasonable in the scenarios where CP makes its pricing decision according to the user response. On the other hand, the scaling factors σ_{eu} and σ_{cp} of EU and CP can be quite different. The sponsoring behavior will be affected by whether the traffic volume is constrained by EU or CP. If $x^*_{eu} \leq x^*_{cp}$, we have $s \leq \frac{\sigma_{cp} p_{eu} - \sigma_{eu} p_{cp}}{(\sigma_{eu} + \sigma_{cp}) p_{eu}}$ and $x^* = x^*_{eu}$. Similarly, if $x^*_{eu} \geq x^*_{cp}$, we have $s \geq \max\left(\frac{\sigma_{cp} p_{eu} - \sigma_{eu} p_{cp}}{(\sigma_{eu} + \sigma_{cp}) p_{eu}}, 0\right)$ and $x^* = x^*_{cp}$. We consider each case.

Case (i) When $x^* = x^*_{cp}$. The profit of the CP can be written as

$$V(s) = \sigma_{cp} \cdot u_{cp}(x^*_{cp}(s)) - s \cdot x^*_{cp}(s) \cdot p_{eu} - x^*_{cp}(s) \cdot p_{cp}. \tag{5.6}$$

By substituting $x^*_{cp}(s, p_{cp}, p_{eu}) = \left(\frac{\sigma_{cp}}{s p_{eu} + p_{cp}}\right)^{\frac{1}{\alpha}}$ into (5.6), it can be easily shown that $V(s)$ is a decreasing function of s, and we have the optimal value $s^* = \max\left(\frac{\sigma_{cp} p_{eu} - \sigma_{eu} p_{cp}}{(\sigma_{eu} + \sigma_{cp}) p_{eu}}, 0\right)$. Thus, the traffic amount and the sponsoring rate will be

$$(x^*, s^*) = (x^*_{cp}, s^*) = \begin{cases} \left(\left(\frac{\sigma_{cp}}{p_{cp}}\right)^{\frac{1}{\alpha}}, \ 0\right), & if \ \ \frac{\sigma_{cp}}{\sigma_{eu}} \leq \frac{p_{cp}}{p_{eu}}, \\ \left(\left(\frac{\sigma_{cp} + \sigma_{eu}}{p_{cp} + p_{eu}}\right)^{\frac{1}{\alpha}}, \ \frac{\sigma_{cp} p_{eu} - \sigma_{eu} p_{cp}}{(\sigma_{eu} + \sigma_{cp}) p_{eu}}\right), & if \ \ \frac{\sigma_{cp}}{\sigma_{eu}} > \frac{p_{cp}}{p_{eu}}. \end{cases} \tag{5.7}$$

The maximum profit of CP is given as

$$V^*(x^*_{cp}, s^*) = \begin{cases} \dfrac{\alpha(\sigma_{cp})^{\frac{1}{\alpha}}}{1 - \alpha}(p_{cp})^{1 - \frac{1}{\alpha}}, & if \ \ \frac{\sigma_{cp}}{\sigma_{eu}} \leq \frac{p_{cp}}{p_{eu}}, \\ \dfrac{\alpha \sigma_{cp}}{1 - \alpha}\left(\dfrac{p_{eu} + p_{cp}}{\sigma_{eu} + \sigma_{cp}}\right)^{1 - \frac{1}{\alpha}}, & if \ \ \frac{\sigma_{cp}}{\sigma_{eu}} > \frac{p_{cp}}{p_{eu}}. \end{cases} \tag{5.8}$$

Case (ii) When $x^* = x^*_{eu}$. In this case, we have $s \leq \frac{\sigma_{cp} p_{eu} - \sigma_{eu} p_{cp}}{(\sigma_{eu} + \sigma_{cp}) p_{eu}}$, $x^*_{eu}(s, p_{eu}) = \left(\frac{\sigma_{eu}}{(1 - s) p_{eu}}\right)^{\frac{1}{\alpha}}$ and $\frac{\sigma_{cp}}{\sigma_{eu}} > \frac{p_{cp}}{p_{eu}}$. CP will optimize its sponsorship percentage by solving

$$\max \quad \frac{\sigma_{cp}\left(\frac{\sigma_{eu}}{p_{eu}}\right)^{\frac{1}{\alpha}-1}}{1-\alpha}(1-s)^{1-\frac{1}{\alpha}} - \frac{(sp_{eu}+p_{cp})\left(\frac{\sigma_{eu}}{p_{eu}}\right)^{\frac{1}{\alpha}}}{(1-s)^{\frac{1}{\alpha}}},$$

$$s.t. \quad 0 \leq s \leq \frac{\sigma_{cp}\,p_{eu}-\sigma_{eu}\,p_{cp}}{(\sigma_{eu}+\sigma_{cp})\,p_{eu}}, \quad \frac{\sigma_{cp}}{\sigma_{eu}} > \frac{p_{cp}}{p_{eu}}. \tag{5.9}$$

From the first order condition, the optimal data rate x^* and the optimal sponsoring rate s^* can be obtained as

$$(x_{eu}^*, s^*) = \begin{cases} \left(\left(\frac{\sigma_{eu}}{p_{eu}}\right)^{\frac{1}{\alpha}}, \quad 0\right), & if \quad \frac{p_{cp}}{p_{eu}} < \frac{\sigma_{cp}}{\sigma_{eu}} \leq \alpha + \frac{p_{cp}}{p_{eu}}, \\ \left(\left(\frac{\sigma_{cp}+(1-\alpha)\sigma_{eu}}{p_{cp}+p_{eu}}\right)^{\frac{1}{\alpha}}, \quad \frac{\frac{\sigma_{cp}}{\sigma_{eu}}-\alpha-\frac{p_{cp}}{p_{eu}}}{\frac{\sigma_{cp}}{\sigma_{eu}}+1-\alpha}\right), & if \quad \frac{\sigma_{cp}}{\sigma_{eu}} > \alpha + \frac{p_{cp}}{p_{eu}}, \end{cases} \tag{5.10}$$

and the maximum profit of CP is

$$V^*(x_{eu}^*, s^*) = \begin{cases} \left(\frac{\sigma_{eu}}{p_{eu}}\right)^{\frac{1}{\alpha}}\left[\frac{\sigma_{cp}\,p_{eu}}{(1-\alpha)\sigma_{eu}} - p_{cp}\right] & if \quad \frac{p_{cp}}{p_{eu}} < \frac{\sigma_{cp}}{\sigma_{eu}} \leq \alpha + \frac{p_{cp}}{p_{eu}}, \\ \frac{\alpha(p_{cp}+p_{eu})}{1-\alpha}\left(\frac{\sigma_{cp}+(1-\alpha)\sigma_{eu}}{p_{cp}+p_{eu}}\right)^{\frac{1}{\alpha}} & if \quad \frac{\sigma_{cp}}{\sigma_{eu}} > \alpha + \frac{p_{cp}}{p_{eu}}. \end{cases} \tag{5.11}$$

From the two-case response of CP, we can obtain the following Proposition.

Proposition 1 *Given prices p_{cp} and p_{eu}, the optimal sponsorship rate s^* of the CP is*

$$case\,(1)\,if \quad \frac{\sigma_{cp}}{\sigma_{eu}} \leq \frac{p_{cp}}{p_{eu}}, \qquad\qquad s^* = 0,$$

$$case\,(2)\,if \quad \frac{p_{cp}}{p_{eu}} < \frac{\sigma_{cp}}{\sigma_{eu}} \leq \alpha + \frac{p_{cp}}{p_{eu}}, \quad s^* = 0, \tag{5.12}$$

$$case\,(3)\,if \quad \frac{\sigma_{cp}}{\sigma_{eu}} > \alpha + \frac{p_{cp}}{p_{eu}}, \qquad s^* = \frac{\frac{\sigma_{cp}}{\sigma_{eu}}-\alpha-\frac{p_{cp}}{p_{eu}}}{\frac{\sigma_{cp}}{\sigma_{eu}}+1-\alpha}.$$

Proof For case 1, the maximum available profit of CP can be easily obtained as $V^*(x_{cp}^*, s^*) = \frac{\alpha(\sigma_{cp})^{\frac{1}{\alpha}}}{1-\alpha}(p_{cp})^{1-\frac{1}{\alpha}}$ from (5.8).

For $\frac{\sigma_{cp}}{\sigma_{eu}} > \frac{p_{cp}}{p_{eu}}$, the CP will choose the largest one among available profits of $V^*(x_{cp}^*, s^*)$ and $V^*(x_{eu}^*, s^*)$, given in (5.8) and (5.11), respectively. Let $\sigma = \frac{\sigma_{cp}}{\sigma_{eu}}$ and $p = \frac{p_{cp}}{p_{eu}}$. We decompose it into two subcases as below.

(1) When $p < \sigma \leq \alpha + p$, each profit function can be written as

$$V^*(x_{cp}^*, s^*) = \frac{(\sigma_{eu})^{\frac{1}{\alpha}}(p_{eu})^{1-\frac{1}{\alpha}}}{(1-\alpha)}\left(\frac{1+p}{1+\sigma}\right)\left(\frac{1+p}{1+\sigma}\right)^{-\frac{1}{\alpha}}\alpha\sigma,$$

$$V^*(x_{eu}^*, s^*) = \frac{(\sigma_{eu})^{\frac{1}{\alpha}}(p_{eu})^{1-\frac{1}{\alpha}}}{(1-\alpha)}(\sigma - (1-\alpha)p).$$

Consider the ratio $\frac{V^*(x^*_{eu},s^*)}{V^*(x^*_{cp},s^*)}$. By using the generalized form of Bernoulli's inequality $(1+x)^r \geq 1 + rx$ for $r \leq 0$ or $r \geq 1$ and $x > -1$, we can obtain

$$\frac{V^*(x^*_{eu},s^*)}{V^*(x^*_{cp},s^*)} \geq \left(\frac{\sigma-(1-\alpha)p}{\alpha\sigma}\right)\left(\frac{1+\sigma}{1+p}\right)\left(1+\frac{p-\sigma}{(1+\sigma)\alpha}\right) = 1 + \frac{(1-\alpha)(\sigma-p)(p+\alpha-\sigma)}{\sigma\alpha^2(1+p)}.$$

Hence, if $p < \sigma \leq \alpha + p$, we have $\frac{V^*(x^*_{eu},s^*)}{V^*(x^*_{cp},s^*)} \geq 1$, implying $x^* = x^*_{eu}$ and $s^* = 0$ from (5.10)

(2) When $\sigma > \alpha + p$, we have

$$V^*(x^*_{cp}, s^*) = \left(\frac{\alpha}{1-\alpha}\right)(p_{eu} + p_{cp})^{1-\frac{1}{\alpha}}(\sigma_{eu})^{\frac{1}{\alpha}}(\sigma)(1+\sigma)^{\frac{1}{\alpha}-1},$$

$$V^*(x^*_{eu}, s^*) = \left(\frac{\alpha}{1-\alpha}\right)(p_{eu} + p_{cp})^{1-\frac{1}{\alpha}}(\sigma_{eu})^{\frac{1}{\alpha}}(1+\sigma-\alpha)^{\frac{1}{\alpha}}.$$

Again we consider the ratio $\frac{V^*(x^*_{eu},s^*)}{V^*(x^*_{cp},s^*)} = \frac{1+\sigma}{\sigma}\left(1-\frac{\alpha}{1+\sigma}\right)^{\frac{1}{\alpha}}$. Applying the generalized form of Bernoulli's inequality, we have $\frac{V^*(x^*_{eu},s^*)}{V^*(x^*_{cp},s^*)} \geq \frac{1+\sigma}{\sigma}\left(1-\frac{1}{1+\sigma}\right) =$ 1, and thus we have $x^* = x^*_{eu}$ and $s^* = \frac{\frac{\sigma_{cp}}{\sigma_{eu}}-\alpha-\frac{p_{cp}}{p_{eu}}}{\frac{\sigma_{cp}}{\sigma_{eu}}+1-\alpha}$ from (5.10).

According to Proposition 1, CP has no incentive to invest in sponsored data plan when $\frac{\sigma_{cp}}{\sigma_{eu}} \leq \alpha + \frac{p_{cp}}{p_{eu}}$. On the other hand, when $\frac{\sigma_{cp}}{\sigma_{eu}} > \alpha + \frac{p_{cp}}{p_{eu}}$, CP will invest in sponsoring as in (5.10). The data rate under sponsoring will be

$$case\ (1)\ if\quad \frac{\sigma_{cp}}{\sigma_{eu}} \leq \frac{p_{cp}}{p_{eu}}, \qquad x^*(p_{cp}, p_{eu}) = \left(\frac{\sigma_{cp}}{p_{cp}}\right)^{\frac{1}{\alpha}},$$

$$case\ (2)\ if\quad \frac{p_{cp}}{p_{eu}} < \frac{\sigma_{cp}}{\sigma_{eu}} \leq \alpha + \frac{p_{cp}}{p_{eu}},\quad x^*(p_{cp}, p_{eu}) = \left(\frac{\sigma_{eu}}{p_{eu}}\right)^{\frac{1}{\alpha}}, \qquad (5.13)$$

$$case\ (3)\ if\quad \frac{\sigma_{cp}}{\sigma_{eu}} > \alpha + \frac{p_{cp}}{p_{eu}},\quad x^*(p_{cp}, p_{eu}) = \left(\frac{\sigma_{cp}+(1-\alpha)\sigma_{eu}}{p_{cp}+p_{eu}}\right)^{\frac{1}{\alpha}}.$$

5.3.2 Utility Maximization of ISP_1

ISP_1 also tries to maximize its total profit in each region specified in (5.13). We obtain the optimal response of ISP_1 in each case.

Case (1) When $x^* = \left(\frac{\sigma_{cp}}{p_{cp}}\right)^{\frac{1}{\alpha}}$ and $s^* = 0$. From (5.3), ISP_1 maximizes $(p_{cp} - p_{tr} - m_1) \cdot \left(\frac{\sigma_{cp}}{p_{cp}}\right)^{\frac{1}{\alpha}}$ subject to $\frac{\sigma_{cp}}{\sigma_{eu}} \cdot p_{eu} \leq p_{cp}$. The best response p^*_{cp} of ISP_1 can be easily obtained as $p^*_{cp} = \frac{p_{tr}+m_1}{1-\alpha}$.

Case (2) When $x^* = \left(\frac{\sigma_{eu}}{p_{eu}}\right)^{\frac{1}{\alpha}}$ and $s^* = 0$. From (5.3), ISP_1 has the objective of

$$\max_{p_{cp} \geq 0} \quad (p_{cp} - p_{tr} - m_1) \cdot \left(\frac{\sigma_{eu}}{p_{eu}}\right)^{\frac{1}{\alpha}} \text{ subject to } \frac{p_{cp}}{p_{eu}} - \frac{\sigma_{cp}}{\sigma_{eu}} \leq 0 \text{ and } \frac{\sigma_{cp}}{\sigma_{eu}} - \alpha - \frac{p_{cp}}{p_{eu}} \leq 0.$$

From the constraints, we have $p_{cp} \in \left[\left(\frac{\sigma_{cp}}{\sigma_{eu}} - \alpha\right) p_{eu}, \frac{\sigma_{cp}}{\sigma_{eu}} p_{eu}\right]$. Note that since the objective is an increasing function of p_{cp}, we set the largest $p_{cp} = \frac{\sigma_{cp}}{\sigma_{eu}} \cdot p_{eu}$ for the optimal solution, which gives us maximum utility $P^* = \left(\frac{\sigma_{cp}}{\sigma_{eu}} \cdot p_{eu} - p_{tr} - m_1\right) \cdot$

$\left(\frac{\sigma_{eu}}{p_{eu}}\right)^{\frac{1}{\alpha}}$. By differentiating it with respect to p_{eu}, we can find $p^*_{eu} = \frac{\sigma_{eu}}{\sigma_{cp}} \cdot \left(\frac{p_{tr}+m_1}{1-\alpha}\right)$ that maximizes P^*, which results in the optimal $p^*_{cp} = \frac{p_{tr}+m_1}{1-\alpha}$.

Case (3) When $x^* = \left(\frac{\sigma_{cp}+(1-\alpha)\sigma_{eu}}{p_{cp}+p_{eu}}\right)^{\frac{1}{\alpha}}$ and $s^* = \frac{\frac{\sigma_{cp}}{\sigma_{eu}} - \alpha - \frac{p_{cp}}{p_{eu}}}{\frac{\sigma_{cp}}{\sigma_{eu}} + 1 - \alpha}$. The problem can

be rewritten as $\max_{p_{cp} \geq 0} (p_{cp} + s^* p_{eu} - p_{tr} - m_1) \cdot \left(\frac{\sigma_{cp}+(1-\alpha)\sigma_{eu}}{p_{cp}+p_{eu}}\right)^{\frac{1}{\alpha}}$, subject to $p_{cp} \leq$

$\left(\frac{\sigma_{cp}}{\sigma_{eu}} - \alpha\right) p_{eu}$. From the first order condition, we can obtain the optimal price $p^*_{cp} = \frac{(k+1)(p_{tr}+m_1)}{k(1-\alpha)} - p_{eu}$, where $k = \frac{\sigma_{cp}}{\sigma_{eu}} - \alpha$.

5.3.3 Utility Maximization of ISP_2

For the behaviors of ISP_2, we also consider the three cases of (5.13) and find the best strategy of ISP_2 for each case.

Case (1) When $x^*(p^*_{cp}, p_{eu}) = \left(\frac{\sigma_{cp}}{p^*_{cp}}\right)^{\frac{1}{\alpha}}$ and $s^* = 0$. We already have $p^*_{cp} = \frac{p_{tr}+m_1}{1-\alpha}$. From (5.4) and (5.13), the ISP_2 determines its prices p_{eu} and p_{tr} by solving

$$\max_{p_{eu} \geq 0, p_{tr} \geq 0} \quad ((1-s^*) \cdot p_{eu} + p_{tr} - m_2) \cdot \left(\frac{\sigma_{cp}}{p^*_{cp}}\right)^{\frac{1}{\alpha}}, \text{ subject to } \frac{\sigma_{cp}}{\sigma_{eu}} - \frac{p^*_{cp}}{p_{eu}} \leq 0.$$

Let P denote the objective function. From the Karush-Kuhn-Tucker (KKT) conditions, we have $\frac{\partial P}{\partial p_{eu}} = 0$, $\frac{\partial P}{\partial p_{tr}} = 0$, and $\lambda \cdot \left[\frac{\sigma_{cp}}{\sigma_{eu}} - \frac{p^*_{cp}}{p_{eu}}\right] = 0$. By solving these equations, we have the optimal prices

$$p^*_{eu} = \frac{(m_1+m_2)}{(1-\alpha)(1+(k+\alpha)(1-\alpha))} \quad and \quad p^*_{tr} = \frac{(k+\alpha)(m_1+m_2)}{(1+(k+\alpha)(1-\alpha))} - m_1, \tag{5.14}$$

at which the maximum profit P^* is $\left[\frac{\alpha(m_1+m_2)}{(1-\alpha)}\right]\left(\frac{\sigma_{cp}(1-\alpha)(1+(k+\alpha)(1-\alpha))}{(k+\alpha)(m_1+m_2)}\right)^{\frac{1}{\alpha}}$, where $k = \frac{\sigma_{cp}}{\sigma_{eu}} - \alpha$.

Case (2) When $x^*(p_{cp}^*, p_{eu}) = \left(\frac{\sigma_{eu}}{p_{eu}}\right)^{\frac{1}{\alpha}}$ and $s^* = 0$. In this case, we have $p_{cp}^* = \frac{p_{tr}+m_1}{1-\alpha}$. From (5.4) and (5.13), the ISP_2 determines its prices by solving the following problem.

$$(ISP_2 - P) \quad \max_{p_{eu} \geq 0, p_{tr} \geq 0} \quad ((1 - s^*) \cdot p_{eu} + p_{tr} - m_2) \cdot \left(\frac{\sigma_{eu}}{p_{eu}}\right)^{\frac{1}{\alpha}},$$

$$s.t. \quad \frac{p_{cp}^*}{p_{eu}} - \frac{\sigma_{cp}}{\sigma_{eu}} \leq 0 \quad and \quad \frac{\sigma_{cp}}{\sigma_{eu}} - \alpha - \frac{p_{cp}^*}{p_{eu}} \leq 0. \tag{5.15}$$

From the KKT conditions, we have $\frac{\partial P}{\partial p_{eu}} = 0$, $\frac{\partial P}{\partial p_{tr}} = 0$, $\lambda_1 \cdot \left(\frac{p_{cp}^*}{p_{eu}} - \frac{\sigma_{cp}}{\sigma_{eu}}\right) = 0$ and $\lambda_2 \cdot \left(\frac{\sigma_{cp}}{\sigma_{eu}} - \alpha - \frac{p_{cp}^*}{p_{eu}}\right) = 0$, where $\lambda_i \geq 0$, $p_{cp} \geq 0$, and $p_{eu} \geq 0$. There are three possible subcases: (i) $\lambda_1 = 0, \lambda_2 \neq 0$, (ii) $\lambda_1 \neq 0, \lambda_2 = 0$, (iii) $\lambda_1 = 0$ and $\lambda_2 = 0$.

(i) When $\lambda_1 = 0$ and $\lambda_2 \neq 0$, the optimal prices will be

$$p_{eu}^* = \frac{m_1+m_2}{(1-\alpha)(1+k(1-\alpha))} \quad and \quad p_{tr}^* = \frac{k(m_1+m_2)}{1+k(1-\alpha)} - m_1, \tag{5.16}$$

where $k = \frac{\sigma_{cp}}{\sigma_{eu}} - \alpha$, and we have the maximum profit $P_{\lambda_1}^* = \left[\frac{\alpha(m_1+m_2)}{(1-\alpha)}\right]\left(\frac{(\sigma_{cp}-\sigma_{eu}\alpha)(1-\alpha)^2+\sigma_{eu}(1-\alpha)}{m_1+m_2}\right)^{\frac{1}{\alpha}}$.

(ii) When $\lambda_1 \neq 0$ and $\lambda_2 = 0$, the optimal prices will be

$$p_{eu}^* = \frac{(m_1+m_2)}{(1-\alpha)(1+(k+\alpha)(1-\alpha))} \quad and \quad p_{tr}^* = \frac{(k+\alpha)(m_1+m_2)}{(1+(k+\alpha)(1-\alpha))} - m_1, \tag{5.17}$$

and the maximum profit $P_{\lambda_2}^* = \left[\frac{\alpha(m_1+m_2)}{(1-\alpha)}\right]\left(\frac{\sigma_{cp}(1-\alpha)^2+\sigma_{eu}(1-\alpha)}{m_1+m_2}\right)^{\frac{1}{\alpha}}$.

(iii) When $\lambda_1 = 0$ and $\lambda_2 = 0$, the two inequality constraints of (5.15) should be an active constraint (i.e., the equalities hold). However, it is not possible to satisfy both equalities, and hence, this case is infeasible.

From $P_{\lambda_2}^* > P_{\lambda_1}^*$, we should have $\lambda_2 = 0$ and the best response of the ISP_2 is (5.17), which equals the result of case 1 in (5.14).

Case (3) In this case, we have the optimal sponsoring rate $s^* = \frac{\frac{\sigma_{cp}}{\sigma_{eu}}-\alpha-\frac{p_{cp}}{p_{eu}}}{\frac{\sigma_{cp}}{\sigma_{eu}}+1-\alpha}$ and the traffic demand is $x^*(p_{cp}^*, p_{eu}) = \left(\frac{\sigma_{cp}+(1-\alpha)\sigma_{eu}}{p_{cp}+p_{eu}}\right)^{\frac{1}{\alpha}}$.

As shown in Sect. 5.3.2, the best-response p_{cp}^* of ISP_1 is $\frac{(k+1)(p_{tr}+m_1)}{k(1-\alpha)} - p_{eu}$. From (5.4) and (5.13), ISP_2 determines its prices by solving $\max_{p_{eu} \geq 0, p_{tr} \geq 0} ((1 - s^*) \cdot$

$p_{eu} + p_{tr} - m_2) \cdot \left(\frac{\sigma_{cp} + (1-\alpha)\sigma_{eu}}{p_{cp}^* + p_{eu}} \right)^{\frac{1}{\alpha}}$, subject to $\frac{p_{cp}^*}{p_{eu}} + \alpha - \frac{\sigma_{cp}}{\sigma_{eu}} \leq 0$. From the KKT conditions, we have $\frac{\partial P}{\partial p_{eu}} = 0$, $\frac{\partial P}{\partial p_{tr}} = 0$, and $\lambda \cdot \left[\frac{p_{cp}^*}{p_{eu}} + \alpha - \frac{\sigma_{cp}}{\sigma_{eu}} \right] = 0$.

By solving the equations, we can obtain without difficulty that

$$p_{eu}^* = \frac{(m_1 + m_2)}{(1-\alpha)(1+k(1-\alpha))} \quad and \quad p_{tr}^* = \frac{k(m_1 + m_2)}{(1+k(1-\alpha))} - m_1. \tag{5.18}$$

The maximum profit P^* will be $\left[\frac{\alpha(m_1 + m_2)}{1-\alpha} \right] \left(\frac{(\sigma_{cp}(1-\alpha) + \sigma_{eu}(1-\alpha)^2)(1+k(1-\alpha))}{(k+1)(m_1 + m_2)} \right)^{\frac{1}{\alpha}}$.

We have shown the optimal responses of the EU, the CP, and two ISPs in a non-cooperative equilibrium. They describe the sponsoring rate s^* and the pricing of p_{cp}^*, p_{eu}^*, and p_{tr}^* when each player maximizes its own utility in a greedy manner.

5.4 Numerical Simulations

We verify our analytical results through numerical simulations. We consider one CP, one EU, and two ISPs, where the CP and the EU share the same utility-level function $\alpha_{eu} = \alpha_{cp} = \alpha \in (0, 1)$. Figure 5.2a shows that CP has incentive to invest in sponsored data plan if $\frac{\sigma_{cp}}{\sigma_{eu}} > \alpha + \frac{p_{cp}}{p_{eu}}$. It means that as CP has higher utility level and EU consuming the content has relatively lower utility level (or similarly, the price charged to CP is relatively lower than the price charged to EU), CP tries to provide a higher sponsorship rate. In contrast, when $\frac{\sigma_{cp}}{\sigma_{eu}} \leq \alpha + \frac{p_{cp}}{p_{eu}}$, CP best strategy is not sponsoring the user access, i.e., $s^* = 0$.

Next we observe the payoff of ISP_2 as we change the price per unit traffic p_{eu} that charges to the user. Figure 5.2b illustrates the results and show that the payoff of ISP_2 linearly rises till some point, and then declines exponentially, which is due to the fact that the demand of users is inversely proportional to p_{eu}. Although ISP_2 obtains its revenue from charging ISP_1 with transit-price p_{tr}, the results show

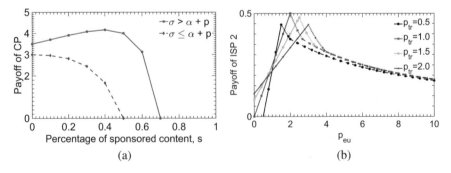

Fig. 5.2 Payoff changes of CP and ISP_2 when $\alpha = 0.5$. (**a**) CP. (**b**) ISP_2 with $\sigma_{cp} = 2$, $\sigma_{eu} = 1$

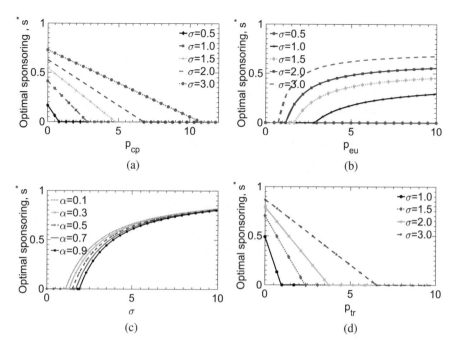

Fig. 5.3 The optimal sponsoring rate with respect to p_{cp}, p_{eu}, σ, and p_{tr}. (**a**) $p_{eu} = 4$, $\alpha = 0.3$. (**b**) $p_{cp} = 2$, $\alpha = 0.3$. (**c**) $p_{cp} = 2$, $p_{eu} = 2$. (**d**) $p_{eu} = 4$, $\alpha = 0.3$

that increasing the p_{tr} does not necessarily increase the payoff of ISP_2. As the transit price becomes higher, CP is forced to increase p_{cp} which in turn results in a decline of the traffic demand. Hence, the maximum point is achieved at $p_{tr} = 1$ and $p_{eu} = 2$.

We now examine the impact of ISP prices (p_{cp}, p_{eu}, and p_{tr}) and σ on the optimal sponsoring rate with different parameter sets. Figure 5.3a shows that as p_{cp} increases, the sponsoring rate drops sharply. The decreasing rate can be mitigated with higher σ. Figure 5.3b shows that with the increase of the p_{eu}, the marginal increase of the sponsoring rate is decreasing. Moreover, a larger σ value indicates a higher and rapidly growing sponsorship rate. Figure 5.3c demonstrates the change of the optimal sponsoring rate with respect to σ under different α values. The sponsorship rate logarithmically increases as σ increases. It can be explained from the fact that the CP with higher revenue level can afford more investment on the sponsoring content. We can also observe that the variation in α has a little impact on the traffic demand. Figure 5.3d will help us to understand the effect of the transit cost p_{tr} to the optimal sponsoring rate s^*. We can observe that the increase of the transit cost results in a sharp drop of s^*. The rise of transit cost will incur significant loss in ISP_1's revenue, which forces ISP_1 to increase its charge to CP, resulting in a rapid drop of the sponsoring rate.

5.5 Conclusion

In this work, we studied the sponsored data and non-cooperative inter-pricing among ISPs that jointly deliver traffic from CPs to EUs. We derived the best response of the EU, the CP, and the ISPs, and analyzed their implications for the sponsoring strategy of the CP. We investigate the interactions between strategic EU, CP, and two interconnected ISPs through a sequential Stackelberg game, and verify our results through numerical simulations. Our results clarify the high impact of the transit price of intermediate ISP on the sponsoring strategies of the CP, and demonstrate in what scenarios sponsoring helps. There are a couple of interesting direction to extend our results. A cooperation between the two ISPs will change the system dynamics and bring a different structure of pricing and sponsoring, and may improve the total payoff of the ISPs at the cost of the EU and the CP. On the other hand, multiple ISPs for the service to the EU or the CP may result in competition between the ISPs and can lead to a higher social welfare.

Acknowledgement This work was in part supported by the NRF grant funded by the Korea government (MSIT) (No. NRF-2017R1E1A1A03070524).

References

1. Developing Telecoms: Data Monetisation Strategies Will Help Telcos Capture Emerging Markets (2014). https://www.developingtelecoms.com/tech/customer-management/7297-data-monetisation-strategies-will-help-telcos-capture-emerging-markets.html
2. Jin, Y., Reiman, M.I., Andrews, M.: Pricing sponsored content in wireless networks with multiple content providers. In: 2015 IEEE Conference on Computer Communications Workshops (INFOCOM WKSHPS), pp. 1–6 (2015)
3. Joe-Wong, C., Ha, S., Chiang, M.: Sponsoring mobile data: an economic analysis of the impact on users and content providers. In: 2015 IEEE Conference on Computer Communications (INFOCOM), pp. 1499–1507 (2015)
4. Lotfi, M.H., Sundaresan, K., Sarkar, S., Khojastepour, M.A.: Economics of quality sponsored data in non-neutral networks. IEEE/ACM Trans. Networking **25**(4), 2068–2081 (2017)
5. Ma, R.T.B.: Subsidization competition: vitalizing the neutral internet. IEEE/ACM Trans. Networking **24**(4), 2563–2576 (2016)
6. Quartz Media: The inside story of how Netflix came to pay Comcast for internet traffic (2017). https://qz.com/256586/the-inside-story-of-how-netflix-came-to-pay-comcast-for-internet-traffic/
7. Sen, S., Joe-Wong, C., Ha, S., Chiang, M.: A survey of smart data pricing: past proposals, current plans, and future trends. ACM Comput. Surv. **46**(2), 15:1–15:37 (2013)
8. Sen, S., Joe-Wong, C., Ha, S., Chiang, M.: Economic Models of Sponsored Content in Wireless Networks with Uncertain Demand, pp. 536. Wiley Telecom (2014)
9. Wu, Y., Kim, H., Hande, P.H., Chiang, M., Tsang, D.H.K.: Revenue sharing among ISPs in two-sided markets. In: 2011 Proceedings IEEE INFOCOM, pp. 596–600 (2011)
10. Xiong, Z., Feng, S., Niyato, D., Wang, P., Zhang, Y.: Economic analysis of network effects on sponsored content: a hierarchical game theoretic approach. In: GLOBECOM 2017 – 2017 IEEE Global Communications Conference, pp. 1–6 (2017)

Chapter 6
Matching Games for 5G Networking Paradigms

S. M. Ahsan Kazmi, Nguyen H. Tran, and Choong Seon Hong

6.1 Introduction

The number of connected network devices in the current cellular network is witnessing an unprecedented growth that is expected to even further grow due to the emergence of novel applications in 5G and beyond networks [8, 21, 45]. In order to serve this tsunami of devices, a number of new networking paradigms have also taken birth [4, 5, 9, 22, 37, 48]. Dense network of small cells coexisting with the traditional networks typically named as heterogeneous networks (HetNets) has witnessed tremendous success in serving these devices [12, 25, 26, 34]. Similarly, non-orthogonal multiple access (NOMA) is also under consideration lately to serve this tsunami of connected devices [20, 30, 41, 44]. Moreover, other notable novel paradigm that is expected to enhance the number of connections economically in the existing networks is wireless network virtualization (WNV) [19, 24, 28, 31, 32, 39]. Furthermore, in order to meet the stringent requirements set by the 5G networks, these paradigms need to coexist leading to a complex heterogeneous multi-tiered network architecture[48]. Resource allocation (RA) in such complex architecture is among one of the biggest challenges and traditional resource allocation approaches based on centralized solution do not apply and fail in such complex networks [5, 38].

S. M. A. Kazmi · C. S. Hong (✉)
Department of Computer Science and Engineering, Kyung Hee University, Yongin, South Korea
e-mail: ahsankazmi@khu.ac.kr; cshong@khu.ac.kr

N. H. Tran
School of Information Technologies, University of Sydney, Sydney, NSW, Australia

Department of Computer Science and Engineering, Kyung Hee University, Yongin-si, Gyeonggi-do, Republic of Korea
e-mail: nguyen.tran@sydney.edu.au

© Springer Nature Switzerland AG 2019
J. B. Song et al. (eds.), *Game Theory for Networking Applications*,
EAI/Springer Innovations in Communication and Computing,
https://doi.org/10.1007/978-3-319-93058-9_6

Centralized resource allocation solutions for such multi-tier network will incur huge message passing to attain global information, higher delays and will be more computationally complex [26]. Thus, these solutions will eventually fail for a dense setting. Therefore, a significant challenge is to design efficient resource allocation approaches that can be implemented in a distributed and self-organizing fashion. Game theory based distributed resource allocation solutions have been recently under consideration as a promising alternative over the traditional optimization based centralized solutions [25, 26, 28, 43, 49]. However, there exist several drawbacks in classical game theoretic based solutions. One major drawback is that each player requires some sort of information from its competing players, thus, significantly increasing the overhead for a dense setting [15].

Matching game is a mathematical framework based on matching theory that has gained recent attention for the resource allocation problems in the field of wireless networks [15, 17]. This is due to its ability to provide distributed solution when considering combinatorial problems which is the case in resource allocation problems. Moreover, other factors to apply matching theory for resource allocation problem are the following [17]:

- It has the capability to capture various wireless communication features.
- It has the ability to model complex environments through preference relations.
- It can be implemented in a distributed fashion and provide low-complexity.

The matching games can be categorized into multiple categories [15, 17, 36]. In this chapter, we focus on two-sided matching game problems as they readily translate to the resource allocation problems for wireless networks. In two-sided matching games, the set of players are typically divided into two distinct non-overlapping sets. Then, the matching problem is defined as to find a match between the players of one set and the players of the corresponding other set, given their individual preferences derived from different objectives. In two-sided matching games, three types of games exist based on the quota values [15, 17, 36, 40]. The first type can be classified as one-to-one matching games. In such games, each player from one set is matched only with a single player of the corresponding set, i.e., the quota is one for both sides. Typical example of this type of matching is the stable marriage problem [14]. The second type corresponds to one-to-many matching games in which a set of players of one side are allowed to match with multiple players (i.e., quota is greater than 1 on one side) of the corresponding side. The college admission problem is an example of these type of games [14]. The final type is the many-to-many classification in which both sides are allowed to match with multiple players of the corresponding sides. The firms and consultants matching problem can be considered as an example of many-to-many matching games [11].

A comprehensive survey on the matching games for future wireless networks has been presented in [15]. Moreover, they also present detailed applications for resource allocation in cognitive radio networks, heterogeneous small cell networks, and device-to-device communications. Similarly, the work in [17] also presents a comprehensive survey on matching games for future wireless networks including other novel paradigms such as Vehicle-to-everything (V2X), millimeter wave communication and LTE-unlicensed. This work can be considered as an extension of these works in which we consider novel paradigms which were not discussed in the aforementioned areas such as dense HetNets, WNV, and NOMA.

The rest of the chapter is organized as follows: In Sect. 6.2, we discuss about the matching theory and its preliminaries. In Sect. 6.3, we investigate the dense heterogeneous paradigm and present its challenges, analytical technique along with its solution concept. In Sect. 6.4, we discuss the details of matching game in the area of wireless network virtualization. Similarly, we present our matching game solution for enabling NOMA in cellular communication in Sect. 6.5. Finally in Sect. 6.6, we conclude our chapter.

6.2 Matching Game Preliminaries

Matching theory is a mathematical framework through which mutual beneficial relations are derived between a set of players. In matching games, the most popular type of games are the two-sided matching games (also known as bipartite matching games) in which the set of players are divided into two sides. Then, each side ranks the corresponding side via a preference relation [15, 17, 36]. The preference relation generally depends upon the local information of a side. Moreover, the players do not require to know other players' preferences and actions to take a decision. The goal is to match players of both sides with each other in the best way. Typical examples include the stable marriage problem in which men and women form the two sides of matching games. Then, both men and women define their preferences over the opposite side based on their local information, i.e., height, color, beauty, intelligence, etc. Finally, the goal is to match men and women with each other in an optimal way [14].

The main goal of the matching game is to design a solution such that the matched players on both sides find the best match. This property is defined as the stability in the context of the matching games and is the key solution concept in this domain [15, 17, 26]. A matching game is stable only if there exists no blocking pair in a matching. A blocking pair is defined as a matched pair (e.g., man, woman pair in case of stable marriage problem) such that it can leave its current matched partner to form a new better pair. The Gale-Shapley (GS) algorithm has been widely used for many matching problems in order to find the stable solution. Note that, in matching games, both sides achieve stability in contrast to the classical game theory Nash equilibrium concept in which only one side stability is generally guaranteed [15].

In wireless communication, the resource allocation problem can be mapped to a two-sided matching game. The two sides here are represented as users and resources. The goal is to match users that form one side to the resources that form the corresponding side in the matching game. Note that the resources in wireless communication can be defined based on the problem scenarios, i.e., physical spectrum, power levels, physical base stations, etc. Similarly, the users can be represented as end user devices, applications, etc. Then, the goal is to define the preference function of each side and rank the corresponding side based on the preference function. Once preferences are defined, we have to design a stable matching scheme for the proposed problem. Next, we present multiple matching games designed for different novel 5G networking paradigms and discuss the challenges and benefits achieved by employing matching theory.

6.3   Matching Game for Dense Heterogeneous Networks

The dense and pervasive deployment of wireless small cells can boost the performance of existing macrocellular networks; however, it poses significant challenges pertaining to the cross-tier interference management. In next-generation 5G networks, the macro base station (MBS) must serve a large coverage area with a high number of connected devices [5]. One important challenge in deploying a large number of small cell base stations (SBSs) is to reduce the signaling and communication overhead at the MBS. A *centralized* RA approach such as in [35, 47] can require the MBS to exchange various information continuously with all SBSs which increases the signaling overhead and computational load for the MBS and becomes impractical for the dense deployment of SBSs. Similarly, a distributed approach with *heavy message exchanges* among network entities over the control channel such as in [2, 42] would again increase signaling and communication overhead. In this work, the downlink resource allocation problem for an underlay small cell network is studied and the protection of the macrocell tier is achieved by imposing cross-tier interference constraints in the resource allocation problem for a dense setting. In a nutshell, a distributed approach with minimum message passing would be more practical and important for the proposed resource allocation problem, because of the reasons mentioned above. To solve this problem, we devise a solution based on matching theory.

6.3.1   System Model

Consider a HetNet consisting of a set of SBSs, $\mathscr{B} = \{1, 2, \ldots, J\}$, located within the coverage of one MBS as shown in Fig. 6.1. The set of macro-cell users (MUEs) and small cell users (SUEs) are denoted by $\mathscr{M} = \{1, 2, \ldots, M\}$ and $\mathscr{S} = \{1, 2, \ldots, S\}$, respectively. The MBS and SBSs use the same set of orthogonal resources $\mathscr{R} = \{1, 2, \ldots, R\}$.[1] However, for any given resource $r \in \mathscr{R}$, a predefined interference threshold $I_{\max}^r$ must be maintained for protecting the MUEs. Moreover, we assume that all SBSs transmit using a fixed power (e.g., any feasible power for

---

[1] One resource corresponds to one subcarrier or subchannel of the LTE network [2].

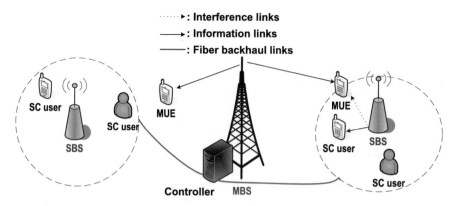

Fig. 6.1 Proposed system model. Solid line showing the downlink information links while dotted line showing the cross tier interference

each SBS transmitter) [42]. However, each SBS can have its own, and different power budgets. In addition, we assume that the transmit power of each SBS is equally divided among its resources and, thus, the interference power on each resource is constant. In this network, our objective is to maximize the sum rate of all SBSs by reusing the macrocell resources.

6.3.2 Problem Statement

Our objective is to maximize the sum rate of all SBSs by reusing the macrocell resources. In order to calculate the sum rate, we need to calculate the received signal to noise ratio (SINR) at each SUE. The received SINR pertaining to the transmission of SBS j to SUE k over resource r with transmit power P_j^r is:

$$\gamma_{j,k}^r = \frac{P_j^r g_{j,k}^r}{P_M^r g_{M,k}^r + \sum_{i \in \Omega_r,} P_i^r g_{i,k}^r + \sigma^2}, \tag{6.1}$$

where P_M^r and $P_i^r, \forall i \in \Omega_r$, represent the transmit powers of the MBS and SBS, respectively, in the set Ω_r which are using resource r. The channel gain between SBS j and SUE k is $g_{j,k}^r$ whereas $g_{M,k}^r$ and $g_{i,k}^r$ are, respectively, the channel gains from the MBS and other underlay SBSs i to SUE k. The noise power is assumed to be σ^2. Then, the data rate of user k associated with SBS j on resource r is given by $R_{j,k}^r = W^r \log(1 + \gamma_{j,k}^r)$ where W^r is the bandwidth of resource r.

Moreover, the interference experienced by MUE m on resource r is given by $I^r = \sum_{j \in \mathcal{B}} \sum_{k \in \mathcal{S}} x_{j,k}^r P_j^r g_{j,m}^r$, where $g_{j,m}^r$ is the channel gain between SBS j and MUE m, on resource r. Note that the binary RA variables $x_{j,k}^r$ ensure that we only account for the interference created by SUEs that are assigned the same resource.

The considered RA problem can be stated as follows:

$$\textbf{RA:} \quad \underset{x_{j,k}^r \in \mathcal{X}, \forall k, j, r}{\text{maximize}} \quad \sum_{j \in \mathcal{B}} R_j \tag{6.2}$$

$$\text{subject to} \quad \sum_{k \in \mathcal{S}} x_{j,k}^r \leq 1, \ \forall r \in \mathcal{R}, \forall j \in \mathcal{B}, \tag{6.3}$$

$$I^r \leq I_{\max}^r, \ \forall r \in \mathcal{R}. \tag{6.4}$$

In **RA**, constraint (6.3) ensures that each resource can be allocated to at most one user in each SBS to avoid strong intra-cell interference; additionally, constraint (6.4) ensures the MUE protection by keeping its aggregate interference below a predefined threshold. Problem **RA** is a non-convex, integer problem, which is difficult to solve for a practical setting with large sets of users and resources [6]. Typically, solutions presented for problems similar to **RA** require significant message exchanges [2, 42]. Therefore, by using matching theory, we present a distributed *novel and practical* algorithm with minimal message passing which is suitable for a large-scale dense networks of HetNets.

6.3.3   *Proposed Solution*

The RA problem can be formulated as a *two-sided matching game*. We assume each SUE can use a single resource from (6.3). However, different SBSs can use the same resource to improve the spectrum efficiency. Our design corresponds to a *many-to-one matching* given by the tuple $(\mathcal{B}, \mathcal{R}, q_r, \succ_{\mathcal{B}}, \succ_{\mathcal{R}})$. Here, $\succ_{\mathcal{B}} \triangleq \{\succ_j\}_{j \in \mathcal{B}}$ and $\succ_{\mathcal{R}} \triangleq \{\succ_r\}_{r \in \mathcal{R}}$ represent the set of the preference relations of the SBSs and resources, respectively.

Definition 1 A *matching* μ is defined by a function from the set $\mathcal{B} \cup \mathcal{R}$ into the setof elements of $\mathcal{B} \cup \mathcal{R}$ such that:

(i) $|\mu(j)| \leq 1$ and $\mu(j) \in \mathcal{R}$,
(ii) $|\mu(r)| \leq q_r$ and $\mu(r) \in \mathcal{B} \cup \phi$, where q_r is the quota of r,
(iii) $\mu(j) = r$ if and only if j is in $\mu(r)$.

6.3.3.1   Preferences of the Players

Matching is performed on the basis of preference profiles that can be built by the SBSs $\mathcal{P}_j$ and the controller $\mathcal{P}_r$ to rank potential matchings based on the local information. Note that, on each r, each SBS j will choose its user k with highest data rate $R_j^r = \max_k R_{j,k}^r$. Then, an SBS j ranks a resource r based on the following preference function:

$$\mathcal{U}_j(r) = R_j^r. \tag{6.5}$$

Similarly, for the controller side, each resource r also ranks the SBSs according to the following preference function:

$$\mathcal{U}_r(j) = R_j^r - \beta I_j^r, \tag{6.6}$$

where $I_j^r = P_j^r g_{j,m}^r$ represent interference produced by SBS j to the MUE assigned that resource. The first term in (6.6) represents the achievable data rate on resource r, the second term accounts for a penalty due to the interference produced by SBS j, and β represents a weight parameter. The second term implies that the controller gives less utility to the SBSs which cause higher interference to the MUE on resource r.

For the formulated two-sided matching game, our goal is to seek a *stable matching*, which is a key solution concept [40]. To find a stable matching, the deferred-acceptance algorithm can be employed [40]. Traditionally, in one-to-many matching, a fixed, per player quota on one side is assumed according to which a fixed number of players of the opposite side can be matched. However, our formulated matching game involves a *dynamic quota* as the controller allows a number of SBSs (with heterogeneous interference) to use each resource as long as the interference constraint on that resource is not violated. This heterogeneous interference of SBSs and dynamic quota of resources introduces new challenges that prevent the use of standard deferred-acceptance algorithm. Therefore, we formally define the blocking pair for the formulated game as follows:

Definition 2 A pair (j, r) is a *blocking pair* for μ if:

a. $I_{res}^r \geq I_j^r$, $j \succ_r \emptyset$ and $r \succ_j \mu(j)$,
b. $I_{res}^r < I_j^r$, $I_{res}^r + \sum_{j' \in \mu(r)} I_{j'}^r \geq I_j^r$, $j \succ_r j'$ and $r \succ_j \mu(j)$,

where $I_{res}^r = I_{max}^r - I^r$ represent the residual of the interference tolerance (remaining quota) on the resource r. *The quota of a resource $r \in \mathcal{R}$ is filled when $I_{res}^r < I_j^r$ for a requesting $j \in \mathcal{B}$.* Definition 2 is based on the following intuition [50]. Whenever an SBS j prefers a resource r to its assigned resource $\mu(j)$, if either: (i) r has sufficient interference tolerance I_{res}^r and is willing to admit j (i.e., $j \succ_r \emptyset$), or (ii) its quota is filled but it is able to admit j by rejecting some accepted SBSs which are ranked lower than j, then j and r can deviate from their assigned $\mu(j)$ and $\mu(r)$, respectively. A matching is stable if no blocking pair exists.

Next, we discuss the "matching" in Definition 1, and then explain the "blocking pair" in Definition 2.

Figure 6.2 shows an example with two set of players, i.e., 3 SBSs and 2 resources. According to Definition 1 (one-to-many matching). A matching μ is an assignment between the two sets of players such that:

1. Each SBS is assigned to at most one resource based on the tolerable interference levels. For example, SBS j_1 is matched to resource r_1. SBS j_2 and SBS j_3 are matched to resource r_2.

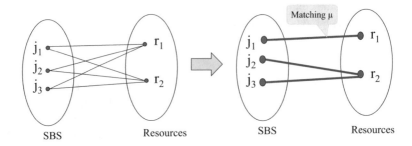

Fig. 6.2 Two-sided matching example

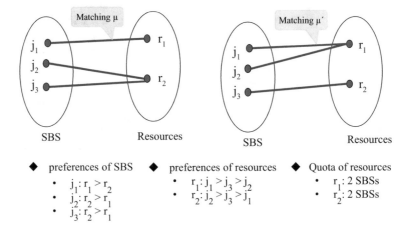

◆ preferences of SBS   ◆ preferences of resources   ◆ Quota of resources
  • $j_1 : r_1 > r_2$ • $r_1 : j_1 > j_3 > j_2$ • r_1: 2 SBSs
  • $j_2 : r_2 > r_1$ • $r_2 : j_2 > j_3 > j_1$ • r_2: 2 SBSs
  • $j_3 : r_2 > r_1$

Fig. 6.3 Two matchings μ and μ'

2. No resource is oversubscribed. For example, assume r_1 and r_2 have limited quota of 1 and 2 players, respectively. Then, r_1 can be matched to one SBS only (j_1) whereas r_2 can support two SBSs, i.e., j_2 and j_3.

Next, we explain the idea of a blocking pair. Figure 6.3 shows an example with two matchings μ and μ'. Additionally, the quota of resources and preference profile for both sides, i.e., SBSs and resources, is provided. Generally, if an SBS j prefers resource r more than its current matched resource, and similarly resource r prefers SBS j more than its current matched SBS. Then, the pair can deviate from the current matching in order to be matched to each other. This is called a blocking pair [40]. In Fig. 6.3, according to the preference profile, it can be seen that the matching produced by μ does not contain any blocking pair. However in μ', SBS j_2 is matched to resource r_1 but SBS j_2 prefers resource r_2 over resource r_1 and similarly resource r_2 also prefers SBS j_2. Thus, a blocking pair exists, i.e., (j_2, r_2) in μ', as both prefer to be matched to each other, over their current partners.

Now, according to Definition 2, if any of the following two condition occurs, then the matching will have a blocking pair:

1. $I_{res}^r \geq I_j^r$, $j \succ_r \emptyset$ and $r \succ_j \mu(j)$.

Whenever an SBS j prefers a resource r to its assigned resource $\mu(j)$ (can be ϕ, meaning it is unassigned), and resource r has sufficient interference tolerance I_{res}^r and also prefers to admit SBS j. Then, SBS j and resource r have a strong incentive to deviate from current matching and form a new matching.

2. $I_{res}^r < I_j^r$, $I_{res}^r + \sum_{j' \in \mu(r)} I_{j'}^r \geq I_j^r$, $j \succ_r j'$ and $r \succ_j \mu(j)$.

Whenever an SBS j prefers a resource r to its assigned resource $\mu(j)$, however, resource r quota is filled but it is able to admit SBS j by rejecting some previously accepted SBSs j' which are ranked lower than SBS j according to the preference profile of resource r. Then, SBS j and resource r can deviate from their assigned matching and block the current matching.

6.3.3.2 Proposed Algorithm

As a solution to this game, we propose a novel RA scheme to produce a stable matching in Algorithm 1 which guarantees macro-tier protection captured in

---

Algorithm 1 Matching-Based Resource Allocation

1: **input**: $\mathcal{P}_j$, $\mathcal{P}_r$, $\forall r, j$

2: **initialize**: $t = 0$, $\mu^{(t)} \triangleq \{\mu(j)^{(t)}, \mu(r)^{(t)}\}_{j \in \mathcal{B}, r \in \mathcal{R}} = \emptyset$, $I_{res}^{r\,(t)} = I_{max}^r$, $\mathcal{K}_r^{(t)} = \emptyset$, $\mathcal{P}_j^{(0)} = \mathcal{P}_j$, $\mathcal{P}_r^{(0)} = \mathcal{P}_r$, $\forall r, j$

3: **repeat**

4:     $t \leftarrow t + 1$

5:     **for** $r \in \mathcal{R}$ **do**

6:         **for** $j \in \mathcal{B}$ with r as its most preferred in $\mathcal{P}_j^{(t)}$ **do**

7:             **while** $j \notin \mu(r)^{(t)}$ and $\mathcal{P}_j^{(t)} \neq \emptyset$ **do**

8:                 **if** $I_{res}^{r\,(t)} \geq I_j^r$, **then**

9:                     $\mu(r)^{(t)} \leftarrow \mu(r)^{(t)} \cup \{j\}$; $I_{res}^{r\,(t)} \leftarrow I_{res}^{r\,(t)} - I_j^r$;

10:                 **else**

11:                     $\mathcal{P'}_r^{(t)} = \{j' \in \mu(r)^{(t)} | j \succ_r j'\}$

12:                     $j_{lp} \leftarrow$ the least preferred $j' \in \mathcal{P'}_r^{(t)}$;

13:                     **while** $(\mathcal{P'}_r^{(t)} \neq \emptyset) \cup (I_{res}^{r\,(t)} < I_j^r)$ **do**

14:                         $\mu(r)^{(t)} \leftarrow \mu(r)^{(t)} \setminus \{j'\}$; $\mathcal{P'}_r^{(t)} \leftarrow \mathcal{P'}_r^{(t)} \setminus \{j_{lp}\}$;

15:                         $I_{res}^{r\,(t)} \leftarrow I_{res}^{r\,(t)} + I_{j'}^r$;

16:                         $j_{lp} \leftarrow$ the least preferred $j' \in \mathcal{P'}_r^{(t)}$;

17:                     **if** $I_{res}^{r\,(t)} \geq I_j^r$, **then**

18:                       $\mu(r)^{(t)} \leftarrow \mu(r)^{(t)} \cup \{j\}$; $I_{res}^{r\,(t)} \leftarrow I_{res}^{r\,(t)} - I_j^r$;

19:                   **else**

20:                       $j_{lp} \leftarrow j$;

21:                   $\mathcal{K}_r^{(t)} = \{k \in \mathcal{P}_r^{(t)} | j_{lp} \succ_r k\} \cup \{j_{lp}\}$

22:                 **for** $k \in \mathcal{K}_r^{(t)}$ **do**

23:                     $\mathcal{P}_k^{(t)} \leftarrow \mathcal{P}_k^{(t)} \setminus \{r\}$; $\mathcal{P}_r^{(t)} \leftarrow \mathcal{P}_r^{(t)} \setminus \{k\}$;

24: **until** $\mu^{(t)} = \mu^{(t-1)}$

25: **output**: $\mu^{(t)}$

---

constraint (5). At each iteration t, each r receives proposals from unassigned SBSs j that rank r as the highest in $\mathscr{P}_j^{(t)}$ (lines 5–7). (i) If r has sufficient quota $I_{res}^r{}^{(t)}$ to admit j, it accepts the proposal and updates $I_{res}^r{}^{(t)}$ and $\mu(r)^{(t)}$ (lines 8–9). (ii) Otherwise, if the quota of r is filled, then r finds all of its current matched j' which have a lower ranking than j according to $\mathscr{P}_r^{(t)}$ (lines 10–11). Each least preferred SBS $j_{lp} \in \mathscr{P}_r'^{(t)}$ is then sequentially rejected, and $I_{res}^r{}^{(t)}$, $\mathscr{P}_r'^{(t)}$, and j_{lp} are updated until j can be admitted or there is no additional j' to reject (lines 12–16). After rejecting all $j' \in \mathscr{P}_r'^{(t)}$, if r still has an insufficient quota to admit j, then j is rejected and j is set to the j_{lp} (lines 17–20). Finally, the controller removes j_{lp} and its less preferred SBSs from the $\mathscr{P}_r^{(t)}$, and similarly these SBSs also remove r from their respective $\mathscr{P}_j^{(t)}$ (lines 21–23). *With this process, we guarantee that any less preferred SBS will not be accepted by that resource even if it has sufficient quota to do so, which is crucial for the matching stability of our design.* This process is repeated until the matching converges (line 24).

Theorem 1 *Algorithm 1 converges to a stable allocation.*

Proof We prove this theorem by contradiction. Assume that Algorithm 1 produces a matching μ with a blocking pair (j, r) by Definition 2. Since $r \succ_j \mu(j)$, j must have proposed to r and has been rejected due to interference violation on r (lines 19–20). When j was rejected, then j' was rejected either before j (lines 13–16), or was made unable to propose because r is removed from j' preference list (lines 22–23). Thus, $j' \notin \mu(r)$, a contradiction.   □

The output $\mu^{(t)}$ of Algorithm 1 can be transformed to a feasible allocation vector $\mathscr{X}$ of problem **RA** (line 25). Note that the worst case running time complexity of Algorithm 1 is *linear* in the size of input preference profiles (i.e., $\mathscr{O}(JR)$ where J and R represent SBSs and resources, respectively).

6.3.3.3 Practical Implementation

To elaborate the practical implementation of matching games for HetNets in detail, we discuss and explain the required overhead that needs to be communicated between both side of players.

1. Initialization phase: Initially, the preference list is set up. SBSs/controller collect information locally, i.e., Interference, propagation gain, transmission rate, monetary offers, etc. However, in our case the maximum interference value on all resource blocks (RBs) is required by each player. Such information can be sent to the players via the Physical Broadcast Channel (PBCH) of LTE. PBCH carries part of the system information, required by the UE terminal in order to access the network, specifically PBCH carries the Master Information Block of 24 bits which has the following information:

   i. DL Bandwidth (3 bits; i.e., 1.4–20 MHz)
   ii. Physical HARQ Indicator Channel (PHICH) Configuration (3 bits)

   iii. System Frame Number (8 bits)
   iv. Spare bits (10 bits)

These spare bits can be used to transmit the maximum interference value required to build the preference profile.

2. Proposing phase: Communication signals are sent from each player to the controller for acquiring resource blocks. This incurs an additional proposal overhead which is not currently supported by the 3GPP standard of LTE. However, this information can be sent via the Physical Uplink control Channel (PUCCH) of each proposing transmitter. Typically, PUCCH carries Uplink Control Information (UCI) which is basically bits and pieces of information that eNB requires from user equipment (UE) in order to understand what UE needs and carries other information like channel quality that UE is observing in downlink. UCI is divided into three main subbranches, i.e. Channel State Information (CSI), Scheduling Requests (SR), and HARQ ACK/NACK. To facilitate these functions, PUCCH has been categorized by seven formats. We can adopt the format 1 and format 1a for sending the proposals. We can see that the proposing message, i.e., scheduling request will be very small, i.e., 1 bit per transmission time interval (TTI).

3. Accepting/rejecting phase: The controller makes decision and sends communication signals with reject/accept overhead. Such information can be sent via Physical Downlink Control Channel (PDCCH) where the set of accepted players will be allocated the RBs and the other will wait. Typically, PDCCH is a physical channel that carries downlink control information (DCI) that carries scheduling assignments and other control information for all network users. There are four formats of PDCCH. The matched lists are then updated by both sides of players. Finally, termination takes place when there is no more RBs to propose or all players are matched with resources. This complete procedure is also shown via the sequence diagram (Fig. 6.4).

6.3.4 Performance Analysis

For our simulations, we consider a network with 5 SBSs each of which supports 3 UEs, and 5 MUEs using 5 resources. All users are randomly located inside the coverage of an MBS which has a radius of $r_1 = 1000$ m, whereas the coverage distance of each small cell is $r_2 = 100$ m. The bandwidth of each resource W^r is set equal to 1 and the weighting parameter β is set to a normalized value of 1, whereas the background noise power is assumed to be -90 dBm. The channel power gain is modeled as $g^r_{j,k} = 10^{(-L(d_{j,k}))/10}$, where $L(d_{j,k})$ represents the path loss and $d_{j,k}$ is the distance between BS j and user k. We assume that $L(d_{M,k}) = 16.62 + 37.6\log_{10}(d_{M,k})$ for the channel gain from the MBS to UE k and $L(d_{j,k}) = 37 + 32\log_{10}(d_{j,k})$ for the channel gain from SBS j to UE k. The SBSs transmit with varying power over simulation runs ranging from 15 to

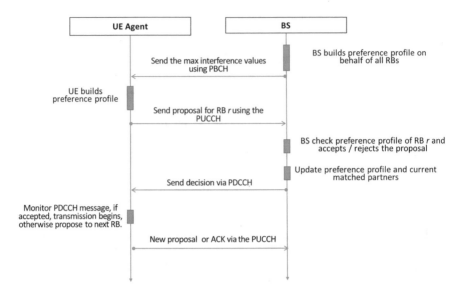

Fig. 6.4 Sequence diagram with communication overhead

23 dBm. For comparison purposes, we compare the proposed matching algorithm with a centralized greedy scheme that sequentially allocates resources to users in each SBS until the interference constraint is violated. Moreover, we also compare the matching based results with an optimization based approach in which the interference threshold (i.e., (6.3)) is equally divided among all SBSs and solved via the dual decomposition approach. All results are obtained by averaging over a large number of independent simulation runs, each of which realizes random locations of base stations, users, and channel power gains. Results corresponding to the optimization-based, matching-based, and greedy algorithms are denoted as "O-DRA," "M-DRA," and "Greedy," respectively.

Figure 6.5 compares the average number of iterations required by both M-DRA and O-DRA versus the number of users (i.e., network size) as $I_{max}^r = -80$ dBm. We can see that, as the number of users increases, the average number of iterations also increases. Moreover, M-DRA has a reasonable convergence time that does not exceed an average of 11 iterations for all network sizes with 5 resources. Moreover, for O-DRA, the maximum number of iterations is smaller than 7 for all network sizes. This fast convergence time can be achieved due to a completely distributed design of O-DRA with no message passing.

In Fig. 6.6, the average sum rate of all UEs versus the number of UEs is shown for the proposed and greedy algorithms as $I_{max}^r = -80$ dBm. Moreover, we use the upper bound (UB) of problem **RA** which is obtained by relaxing the binary indicator variable so that it can take any value in the range [0, 1] as a benchmark here. It can be inferred that the matching-based, optimization-based, and Greedy approaches achieve up to 96.8, 82.6, and 80.2% of the average sum rate obtained by

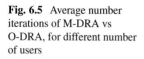

Fig. 6.5 Average number iterations of M-DRA vs O-DRA, for different number of users

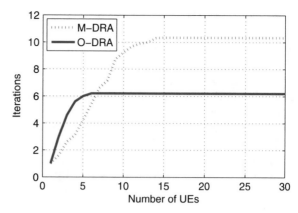

Fig. 6.6 Comparison of average sum rate of O-DRA, M-DRA, and Greedy with UB

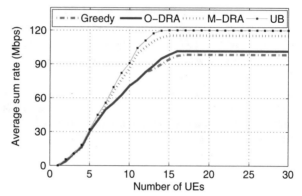

the UB, respectively, for a network with 20 UEs. Thus, it is clear that the matching-based approach is close to optimal. Furthermore, it can be observed that the sum rate increases with more UEs, which, however, saturates as the number of UEs becomes sufficiently large. This is because of the limited number of resources at each SBS ($r = 5$). Additionally, the optimization-based approach achieves a performance benefit up to 4% compared to the greedy approach while the matching-based approach achieves 17 and 21% higher sum rate compared to the optimization-based and greedy approaches, respectively, for a network with 20 UEs.

6.4 Matching Game for Wireless Network Virtualization

Wireless network virtualization (WNV) is a promising candidate to support the deluge of cellular traffic for the forthcoming fifth generation (5G) networks [32]. In a WNV, infrastructure providers (InPs) provide their physical resources as a service to the mobile virtual network operators (MVNOs) to serve its users. The physical resources (i.e., spectrum, power, backhaul/fronthaul, and antennas) of an InP are abstracted into isolated virtual resources (i.e., slices) which are then transparently shared among different MVNOs. Efficient allocation of physical resources to end users has received significant attention in a single-cell WNV scenario [31]. However, a practical deployment of a WNV involves a multi-cell scenario where the coverage area of a specific region will be serviced by a set of InPs. Then, a significant challenge pertaining to such a scenario is the efficient allocation of the resources such that the total performance of WNV over a specific region is improved. Moreover, traditional resource allocation approaches based on single-cell WNV do not directly apply to multi-cell WNV.

Typically resource allocation in WNV can be implemented either by directly allocating resources from an InP-BS to MVNO users (see the works [39, 46] and the reference therein for such approaches) or allocating resources from an InP to an MVNO that further decides the allocation for its users, i.e., works in [33, 53] and the reference therein. This work focuses on the latter approach which makes the resource allocation problem a hierarchical (i.e., two-level) problem [51, 52]. Then, to address this two-level resource allocation problem, we design a hierarchical matching mechanism which can solve the hierarchical (i.e., two-level) WNV problem. In our model, first, service selection is performed in which users are associated to the MVNOs and then each MVNO is provided slices from InPs to serve its users. By adopting such a model, the computational load of an InPs is reduced because now InP is only responsible for allocating resources to each MVNO compared to existing works [39] where the resource allocation has to be obtained directly for all users.

6.4.1 System Model

Consider a downlink of a cellular network consisting of a set of N base stations (BSs), each representing a cell which is owned by an InP.[2] The InP provides its virtual network service to a set of M MVNOs by individual contracts. Moreover, an MVNO $m \in M$ provides its service to a set K_m of subscribed UEs. Then, $K = \cup_m K_m$ represent the total number of UEs. We use notation $|K|$ to denote the cardinality of a set K. Figure 6.7 illustrates our system model.

6.4.2 Channel Model and Assumptions

Each InP owns a set of C_n orthogonal channels, each with bandwidth W. We consider a system with static inter-InP interference such that the interference from other InPs is absorbed into the background noise σ^2. Note that in this work, we assume that all InPs own orthogonal channels, i.e., no interference between InPs allocation.

---

[2]InPs belong to different vendors that own orthogonal frequency channels through administrative licensing.

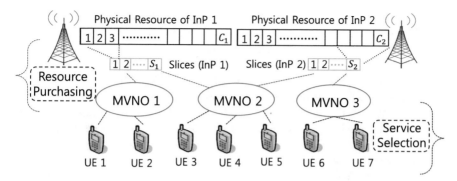

Fig. 6.7 System model: the InP owns the physical resources, virtualizes them into slices, and allocates to multiple MVNOs

The motivation for such an assumption is based on the fact that in our scenario, the InP-BSs belong to different vendors that always own orthogonal frequency channels through administrative licensing (e.g., in the USA, the Federal Communications Commission determines the specified spectrum portions, as well as the vendors who will have access to them). Moreover the current work does not consider the case such as in some existing works [13, 23] in which a single vendor owns multiple BSs and spectrum reuse is applied for spectrum efficiency. Furthermore, we assume equal power on every channel of an InP n, i.e., $P_n = \frac{P_n^{\max}}{|C_n|}$, where P_n is the power on each channel and $P_n^{\max}$ is the maximum power of an InP n. Moreover, InP n provides isolated services by a set of S_n slices, where each slice s_n allocated by InP n to MVNOs m will include heterogeneous number of channels based on MVNO's m demand. Then, the data rate for a UE k on a slice s_n of an InP n is:

$$R_{n,k}^{s_n} = \sum_{c \in s_n} W \log \left(1 + \gamma_{n,k}^c \right), \qquad (6.7)$$

where $\gamma_{n,k}^c = \frac{P_n g_{n,k}^c}{\sigma^2}$, $g_{n,k}^c$ represents the channel gain between InP-BS n and UE k on channel c of slice s_n.

6.4.3 Problem Formulation

The goal in WNV is to maximize the objectives of all UEs, MVNOs, and InPs. Each UE $k \in K$ chooses its service as follows:

$$\text{UE}: \min_{x_{k,m} \in \{0,1\}} \sum_{m \in M} x_{k,m} \beta_m^M d_k, \qquad (6.8)$$

$$\text{s.t.} \sum_{m \in M} x_{k,m} = 1, \qquad (6.9)$$

where $x_{k,m} \in X$ is the binary variable with $x_{k,m} = 1$ indicating that UE k proposes to MVNO m for service selection and $x_{k,m} = 0$ otherwise, d_k represents the demand of the UE k, and β_m^M represents the per unit price of MVNO m. Minimizing (6.8) achieves the UE's goal to pay the minimum for its demand and the constraint in (6.9) represents that a UE can be serviced by only one MVNO. Moreover, we consider the assumptions for constant pricing and constant demand in this work. Typically the solutions (i.e., distributed) presented for joint pricing and allocation problems (i.e., dynamic pricing and demands) assume that the price and demand is kept fixed during the allocation phase and can be updated in the pricing phase. Such an assumption is aligned for a number of joint allocation problems [13]. In our work, by stating the assumption that the price and demand is fixed, we mean that these assumptions hold during the allocation phase only. Moreover, it is certainly of interest, for future work, to design an algorithm that jointly considers the pricing and allocation for WNV. Thus, the current work would constitute a key building block for such future framework. Next, each MVNO m aims to serve its UEs by determining the required channels with the least cost and optimizing its bandwidth according to the slice price offered by the InP. Then, the MVNO m problem is given as:

$$\textbf{MVNO}: \max_{\tilde{x}_{k,m}, \tilde{y}_{m,n}^{s_n} \in \{0,1\}} \sum_{k \in K} \tilde{x}_{k,m} \beta_m^M d_k$$

$$- \sum_{n \in N} \sum_{s_n \in S_n} \tilde{y}_{m,n}^{s_n} \beta_n^I |s_n|, \tag{6.10}$$

$$\text{s.t.} \sum_{m \in M} \tilde{x}_{k,m} \leq 1, \ \forall k, \tag{6.11}$$

$$\sum_{k \in K} \tilde{x}_{k,m} l_{k,n} \leq \tilde{y}_{m,n}^{s_n} |s_n|, \ \forall n, \tag{6.12}$$

where $\tilde{x}_{k,m} \in \tilde{X}$ is the binary service selection decision variable with $\tilde{x}_{k,m} = 1$ indicating that UE k proposal is accepted by MVNO m, $\tilde{y}_{m,n}^s \in \tilde{Y}$ is the binary variable with $\tilde{y}_{m,n}^{s_n} = 1$ denoting that MVNO m proposes to buy slice s_n of InP n and $\tilde{y}_{m,n}^{s_n} = 0$ otherwise. β_n^I is the InP n's per unit price, and $l_{k,n}$ is the required channels to fulfill d_k on InPs n which is calculated by MVNO (details in Sect. 6.4.4.1). Moreover, (6.11) ensures that k is serviced by at most one MVNO and (6.12) ensures that the allocated resources on slice are less than the capacity of slice provided to a MVNO m by InP n.

Finally, the InP aims to satisfy the demands of MVNOs such that the contracts agreements are not violated by solving the following:

$$\textbf{InP}: \max_{y_{m,n}^{s_n} \in \{0,1\}} \sum_{m \in M} \sum_{s_n \in S_n} y_{m,n}^{s_n} \left(\sum_{k \in K_m} \log(R_{n,k}^{s_n}) + \omega \beta_n^I |s_n| \right) \tag{6.13}$$

$$\text{s.t.} \sum_{m \in M} \sum_{s_n \in S_n} y_{m,n}^{s_n} \leq |S_n|, \tag{6.14}$$

$$\sum_{k \in K_m} \sum_{s_n \in S_n} y_{m,n}^{s_n} R_{n,k}^{s_n} \geq d_m, \ \forall m, \tag{6.15}$$

where $y^{s_n}_{m,n} \in Y$ is the binary resource purchasing decision of InP with $y^{s_n}_{m,n} = 1$ indicating that InP n accepts the slice s_n buying proposal of MVNO m and d_m represents the UE demand of the MVNO m (i.e., $d_m = \sum_{k \in K_m} d_k$). The objective function in (6.13) represents the proportional fairness among UEs' [29] and InP revenue in the first and second term, respectively. ω is a weight characterizing the trade-off between fairness and InP's revenue. One of the fundamental requirements of WNV is the isolation among different MVNOs which is achieved by guaranteeing certain predetermined requirements or contract service agreement (e.g., minimum share of resource or data rate) by the InP. We consider isolation at the physical resource level, i.e., channels [53]. Through (6.14) we ensure that allocated slices are less than total slices owned by an InP and the constraint in (6.15) ensures the contract agreement that is considered as an isolation constraint.

Unfortunately, the optimization problem that optimizes the objectives of all UEs, MVNOs, and InPs is a mix integer linear programming problem, which is NP-hard due to its combinatorial nature [6]. Obtaining a central optimal solution (e.g., using exhaustive search) for this problem incurs: (i) heavy computational workload, and (ii) privacy issues between UEs, MVNOs, and InPs. Therefore, by using matching theory which has the ability to solve combinatorial problems [40], we present a distributed approach. Our approach consists of two-level matchings that is able to find a suboptimal solution without any third party rule-enforcing authority and achieve lower-complexity.

6.4.4 Proposed Solution

Our aim in the high-level is to allocate slices from InPs to MVNOs such that the isolation among slices is maintained and maximum demand for each MVNO is fulfilled. Note that we assume a dense network, thus, all MVNO demands cannot be met. However, through our approach (in the high-level), each InP chooses a set of MVNO's demand to serve which maximizes its objective, i.e., maximizes the InP revenue while achieving proportional fairness among the UEs. Similarly, in low-level, the aim is to choose the set of UEs for each MVNO. Moreover, we do not restrict an MVNO to buy slices from only one InP. Then, the MVNO can select a set of UEs for each InP such that the demands of these UEs are less than the available resources of each InP. Furthermore, an MVNO accepts a set of UE such that it can maximize its revenue.

The proposed hierarchical matching game consists of two levels in which matching between UE and MVNO is performed in the low-level while matching between MVNO and InP is at high-level as shown in (Fig. 6.8). Both matching problems can be formulated as a two-sided matching game. Specifically, in the high-level, the InP, who owns the physical resources, acts as the vendor and the MVNOs act as the buyer. In the low-level, each MVNO plays the vendor role and the UEs act as the buyers. It is assumed that each buyer can be associated to only

Fig. 6.8 Block diagram of
hierarchical matching scheme

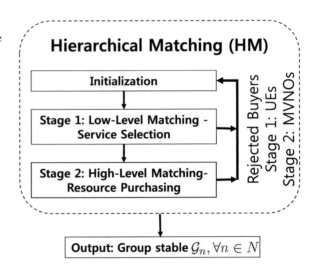

one vendor. However, a vendor can accommodate multiple buyers. Thus, our design
corresponds to a *many-to-one matching* given by the tuple $(B, V, q_v, \succ_B, \succ_V)$.
Here, $\succ_B \triangleq \{\succ_b\}_{b \in B}$ and $\succ_V \triangleq \{\succ_v\}_{v \in V}$ represent the set of the preference
relations of the buyers B and vendors V, respectively.

Definition 3 A *matching* μ is defined by a function from the set $B \cup V$ into the set
of elements of $B \cup V$ such that:

(i) $|\mu(b)| \leq 1$ and $\mu(b) \in V$,
(ii) $|\mu(v)| \leq q_v$ and $\mu(v) \in 2^{|B|} \cup \phi$, where q_v is the quota of v,
(iii) $\mu(b) = v$ if and only if b is in $\mu(v)$.

6.4.4.1 Low-Level Matching Game Between MVNO and UE

In our model, we assume that the available channels and price per channel are
broad-casted in the network by each InP before the matching procedure starts. Thus
each MVNO attains the knowledge of available channels in the network by each
InP. Moreover, as stated above that each MVNO can buy resources from multiple
InPs. Therefore, inspired by the works in [7], for each MVNO m, we create i
dummies ($i \in N$ represents the InP-BS) where the quota of each dummy MVNO is
represented by q_{m_n}. Therefore, for clear presentation in the revised manuscript, we
have modified the MVNO (seller) preference profile representation to $\mathscr{P}_{m_n}^l$, where
$\mathscr{P}_{m_n}^l$ represents the dummy MVNO's m_n preference profile with quota q_{m_n}. Then,
we consider each UE k builds a preference profile for each dummy MVNO m_n
using (6.16) as follows:

$$U_k^{b_l}(m_n) = \beta_{m_n}^M d_k, \quad \forall m_n. \tag{6.16}$$

Note that as the price of each MVNO m with i dummies will be same, the UEs will have the same ranking for these dummy MVNOs. Therefore, to achieve a strict ranking for each dummy MVNO m_n in the preference profile of each UE k, we simply break the tie among these dummy MVNOs m_n by a random selection [27]. Similarly, each dummy MVNO m_n also ranks all UEs based on the profit they yield through (6.17) as follows:

$$U_{m_n}(k) = \max(\beta^M_{m_n} d_k - \beta^I_n l_{k,n}, 0), \ \forall k. \tag{6.17}$$

To rank a UE k, the dummy MVNO m_n needs to calculate the value of $l_{k,n}$ (i.e., required number of channels) to full-fill d_k. Specifically, all UEs are ranked based on the profit they yield in a non-increasing order in the preference profile $\mathscr{P}^l_{m_n}$. In the remainder of this work, we omit the term "dummy" without confusion. Note that here, a UE k is assumed to be indifferent towards all the channels provided by a single InP-BS n because of homogeneous channel gain values (i.e., the channel gain values of different channels owned by an InP-BS are the same for a UE k, while they can be different for different InP-BSs). This assumption is practical as channel gains in current cellular systems (i.e., 4G) are computed by dividing the available frequency band in sub-bands, thus, having a channel gain value for a group of channels [27]. Furthermore, if the revenue from a UE k is negative, that UE is not ranked in $\mathscr{P}^l_{m_n}$ by the MVNO. However from (6.12), each MVNO can only serve limited UEs, i.e., $q_{m_n}{}^3$ which is upper bounded by the slice provided to it by the InP. Then, the goal is service selection of each UE k to an MVNO m_n via matching.

6.4.4.2 High-Level Matching Game Between MVNO and InP

Once a solution to the low-level matching game is obtained, we can solve the high-level game. Here, each MVNO (i.e., dummy MVNO) requires a slice from a specific InP to serve the UEs matched to it in the low-level stage. We denote the demand of each MVNO as $d_{m_n} = \sum_{k \in \mu(m_n)} d_k$. Now both MVNOs and InPs define their respective preference profiles as $\mathscr{P}^u_{m_n}$ and $\mathscr{P}_n$. Here, an MVNO targets to reduce its cost to obtain a slice using (6.18) as follows:

$$U^{b_h}_{m_n}(n) = \beta^I_n d_{m_n}, \ \forall n. \tag{6.18}$$

For the InPs, through (6.13), the goal is to maximize its revenue while maintaining proportional fairness among the UEs. Therefore, it ranks the buyers in a non-increasing manner:

$$U^{s_h}_n(m_n) = \sum_{k \in \mu(m_n)} \log(R^{s_n}_{n,k}) + \omega \beta^I_n \gamma_{m_n}, \ \forall m_n. \tag{6.19}$$

$^3 q_{m_n}$ represents the available channels of InP-BS n.

Here, we assume that the values of d_{m_n} and the set of UEs that are matched in the low-level stage (i.e., $k \in \mu(m_n)$) are sent to the InPs in the proposal phase. Then, InP calculates the required slice size, i.e., γ_{m_n} to fulfill MVNO's m_n demand and γ_{m_n} represents the number of channels in a slice, i.e., $|s_n|$. This information is required by the InP to rank an MVNO m_n through (6.19). Once this information is acquired, each InP can rank all the MVNOs.

6.4.4.3 Hierarchical Game Challenges

For the two-sided hierarchical matching game, our goal is to seek a stable matching, which is a key solution concept in matching theory [40]. To find a stable matching, the deferred-acceptance algorithm can be employed. However, our formulated game involves a *hierarchal* structure and *heterogeneous demands* of buyers. Due to heterogeneous demands, a vendor allows variable numbers of buyers until its quota constraint is not violated [26]. These aforementioned challenges prevent the use of standard deferred-acceptance algorithm. Therefore, we formally define the blocking pair for the formulated game as follows:

Definition 4 A matching μ is *stable* if there exists no blocking pair $(A', v) \in 2^{|B|} \cup V$ with $A' \neq \phi$, such that $v \succ_b \mu(b)$, $\forall b \in A'$ and $(A \cup A') \succ_v \mu(v)$, $A \subseteq \mu(v)$, where $\mu(b)$ and $\mu(v)$ represent, respectively, the current matched partners of vendors and buyers.

Definition 4 is based on the following intuition, a pair (A', v) blocks a matching μ, if vendor v is willing to accept the buyers in A', possibly after rejecting some of its current buyers in $\mu(v)$, i.e., $A \subseteq \mu(v)$ and all buyers $b \in B$ prefer v over their current match $\mu(b)$. In our game, a stable solution ensures that no matched vendor v would benefit from deviating from their assigned buyers b with a new buyer b'. To tackle this challenge, we propose a novel stable matching algorithm in Algorithm 2. The algorithm has two stages, namely, the *Low-Level Matching-Service Selection* stage and the *High-Level Matching- Resource Purchasing* stage. However, Definition 4 is not enough for stating the stability for our proposal as our game involves a *hierarchal* structure. In hierarchical games, a change in player's strategy at a low-level will cause changes in strategy set of players at higher level and, thus, the convergence cannot be achieved until the strategy set of players at low-level is fixed. Therefore, to find a stable solution, we have to guarantee that no change in players' strategy occurs at the low-level once convergence is achieved [16, 18]. We address this challenge by creating a group $\mathscr{G}_n$ for each InP n which is formed as a result of both low-level (i.e., $\mu(m_n)$) and high-level (i.e., $\mu(n)$) stages. Formally, we define the group stability as:

Definition 5 The group $\mathscr{G}_n$, $\forall n \in N$ is said to be stable if it is not blocked by any group $\mathscr{G}_n'$ which is represented by the following two conditions:

(i) No UE k outside the group $\mathscr{G}_n$ can join it.
(ii) No UE k inside the group $\mathscr{G}_n$ can leave it.

Algorithm 2 Hierarchal matching algorithm (HM)

1: **initialize:** $\tau = 0, \mathscr{G}_n^\tau = 0, \forall n$.
2: **while** $\mathscr{G}_n^\tau \neq \mathscr{G}_n^{\tau+1}$ **do**
3: $\quad \tau = \tau + 1$.
  ***Stage 1: Low-Level Matching - Service Selection*:**
4: $\quad$ **input:** $t = 0, q_{m_n}^{(0)} = q_{m_n}^\tau, \mathscr{P}_k^{(0)} = \mathscr{P}_k, \mathscr{P}_{m_n}^{(0)} = \mathscr{P}_{m_n}^l, \forall m_n, k \notin \mathscr{G}_n^\tau$.
5: $\quad t \leftarrow t + 1, \forall k \in \mathscr{K}$, propose to m_n according to $\mathscr{P}_k^{(t)}$.
6: $\quad$ **while** $k \notin \mu(m_n)^{(t)}$ and $\mathscr{P}_k^{(t)} \neq \emptyset$ **do**
7: $\quad\quad$ **if** $q_{m_n}^{(t)} \leq l_{k,n}$ **then**
8: $\quad\quad\quad \mathscr{P}'^{(t)}_{m_n} = \{k' \in \mu(m_n)^{(t)} | k \succ_{m_n} k'\} \cup \{k\}$.
9: $\quad\quad\quad k'_{lp} \leftarrow$ the least preferred $k' \in \mathscr{P}'^{(t)}_{m_n}$.
10: $\quad\quad\quad$ **while** $(\mathscr{P}'^{(t)}_{m_n} \neq \emptyset) \cup (q_{m_n}^{(t)} \geq l_{k,n})$ **do**
11: $\quad\quad\quad\quad \mu(m_n)^{(t)} \leftarrow \mu(m_n)^{(t)} \setminus k'_{lp}, \quad \mathscr{P}'^{(t)}_{m_n} \leftarrow \mathscr{P}'^{(t)}_{m_n} \setminus k'_{lp}$.
12: $\quad\quad\quad\quad q_{m_n}^{(t)} \leftarrow q_{m_n}^{(t)} + l_{k'_{lp},n}, \quad k'_{lp} \leftarrow k' \in \mathscr{P}'^{(t)}_{m_n}$.
13: $\quad\quad\quad$ Remove rejected players from $\mathscr{P}_k^{(t)}$ and $\mathscr{P}_{m_n}^{(t)}$.
14: $\quad\quad$ **else**
15: $\quad\quad\quad \mu(m_n)^{(t)} \leftarrow \mu(m_n)^{(t)} \cup \{k\}, \quad q_{m_n}^{(t)} \leftarrow q_{m_n}^{(t)} - l_{k,n}$.
16: $\quad \tilde{X} \leftarrow \mu^*$
  ***Stage 2: High-Level Matching- Resource Purchasing*:**
17: $\quad$ **input:** $t = 0, q_n^{(0)} = q_n^\tau, \mathscr{P}_{m_n}^{(0)} = \mathscr{P}_{m_n}^u, \mathscr{P}_n^{(0)} = \mathscr{P}_n, \forall m_n, n$.
18: $\quad t \leftarrow t + 1, \forall m_n$, propose to n according to $\mathscr{P}_{m_n}^{(t)}$.
19: $\quad$ **while** $m_n \notin \mu(n)^{(t)}$ and $\mathscr{P}_{m_n}^{(t)} \neq \emptyset$ **do**
20: $\quad\quad$ **if** $q_n^{(t)} \leq |\gamma_{m_n}|$ **then**
21: $\quad\quad\quad \mathscr{P}'^{(t)}_n = \{m_n' \in \mu(n)^{(t)} | m_n \succ_n m_n'\} \cup \{m_n\}$.
22: $\quad\quad\quad m_n'{}_{lp} \leftarrow$ the least preferred $m_n' \in \mathscr{P}'^{(t)}_n$.
23: $\quad\quad\quad$ **while** $(\mathscr{P}'^{(t)}_n \neq \emptyset) \cup (q_n^{(t)} \geq |\gamma_{m_n}|)$ **do**
24: $\quad\quad\quad\quad \mu(n)^{(t)} \leftarrow \mu(n)^{(t)} \setminus m_n', \quad \mathscr{P}'^{(t)}_n \leftarrow \mathscr{P}'^{(t)}_n \setminus m_n'{}_{lp}$.
25: $\quad\quad\quad\quad q_n^{(t)} \leftarrow q_n^{(t)} + |\gamma_{m_n'{}_{lp}}|, \quad m_n'{}_{lp} \leftarrow m_n' \in \mathscr{P}'^{(t)}_n$.
26: $\quad\quad\quad$ Remove rejected players from $\mathscr{P}_{m_n}^{(t)}$ and $\mathscr{P}_n^{(t)}$.
27: $\quad\quad$ **else**
28: $\quad\quad\quad \mu(n)^{(t)} \leftarrow \mu(n)^{(t)} \cup \{m_n\}, \quad q_n^{(t)} \leftarrow q_n^{(t)} - |\gamma_{m_n}|$.
29: $\quad Y \leftarrow \mu^*$
30: $\quad$ Update $\mathscr{G}_n^\tau, \forall n$.
31: **output:** Convergence to group stable $\mathscr{G}_n, \forall n$.

6.4.4.4 Proposed Algorithm

After initialization (line 1), all UEs that do not belong to any group $\mathscr{G}_n^\tau$ join the low-level stage and build the preference profiles for iteration τ. Then, each unassigned UE k proposes to its most preferred MVNO m_n according to $\mathscr{P}_k$ (lines 5–6). (i) If MVNO m_n quota is full, then it finds the current matched UEs k' that ranks lower than k in its preference profile, i.e., $\mathscr{P}'^{(t)}_{m_n}$. Each least preferred UE $k'_{lp} \in \mathscr{P}'_{m_n}t)$ is then sequentially rejected until k can be admitted or there is no additional k' to reject (lines 7–12). If MVNO m_n still has insufficient quota to admit k, then k is also rejected. All rejected UEs and MVNOs then update there respective preference profiles by deleting the rejected players (line 13). (ii) Otherwise, k is accepted and the MVNO m_n updates its quota (lines 14–15). This process is carried out iteratively until either all the UEs are assigned to MVNOs or there are no more MVNOs to propose. This stage terminates when the outcome of two consecutive stage iterations t remains unchanged [40]. The output of this stage μ^* can be transformed to a feasible service selection vector $\tilde{X}$. After the low level matching, MVNOs and InPs build their respective preference profiles based on the output of low-level matching.

Similar to stage 1, after building the preference profile of both sides, each unassigned MVNO m_n proposes to its most preferred InP n according to $\mathscr{P}_{m_n}^{b_h}$. (i) If InP n quota is full, then it finds the current matched MVNOs m_n' that ranks lower than m_n in its preference profile, i.e., $\mathscr{P}_n'^{(t)}$. Each least preferred MVNO $m_{n_{lp}}' \in \mathscr{P}_n'^{(t)}$ is then sequentially rejected until m_n can be admitted or there is no additional m_n' to reject. If InP n still has insufficient quota to admit m_n, then m_n is also rejected. All rejected MVNOs and InPs then update their respective preference profiles by deleting the rejected players. (ii) Otherwise, m_n is accepted and the InP updates its quota. This process is carried out iteratively until either all the MVNOs are assigned to InPs or there are no more InPs to propose. Then, the outcome of this stage can be transformed to a feasible resource purchasing vector between MVNOs and InPs, i.e., Y for high-level matching. Each InP n then updates the group $\mathscr{G}_n = \{(k, m_n) | k \in \mu(m_n), m_n \in \mu(n), \forall k, \forall m_n\}$ which constitutes both the accepted UEs and MVNOs at both the low- and high-level stages. After completion of both these stages, the rejected buyers in both stages, i.e., UE k in low-level and MVNO m_n (consists of UEs that were accepted in low-level by MVNO m_n) will enter into the next iteration $\tau + 1$ as new UEs. Then, both these stages will be executed again with updated values of remaining InP and MVNO quotas. The algorithm terminates once the groups $G_n, \forall n$ do not change for two consecutive iterations τ. This means that there are no further requests from UEs or the available quota is not enough to fulfill any further UEs' request. Finally we claim that the revised Algorithm 2 will converge when there are no new UEs that can enter or leave the stable group $\mathscr{G}_n, \forall n \in N$. Now based on the presented definition of group stability, we claim that Algorithm 2 converges to group stability. Formally we state this as follows:

Theorem 2 *Algorithm 2 converges to group stable output* $\mathscr{G}_n, \forall n \in N$.

Proof Consider a stable group $\mathscr{G}_n$ of (k, m_n) pairs that is formed at the end of the iteration τ. Assuming there exists a UE buyer k' such that $k' \notin \mathscr{G}_n, k' \neq k$. Then, for condition (i) of Definition 1, there can exist two cases.

Case 1 This case states that both UEs k and k' are selected by an MVNO m_n in the low-level stage, i.e., k' and $k \in \mu(m_n)$. Since $m_n \in \mathscr{G}_n$, then all members of m_n also belong to $\mathscr{G}_n$. Thus, $(k', m_n) \in \mathscr{G}_n$ which implies $k' = k$.

Case 2 This case states that there exist a $k' \in \mu(m'_n)$ and $k \in \mu(m_n)$. As $k' \notin \mathcal{G}_n$, then m'_n is rejected at iteration τ as InP n has higher preference for MVNO m_n compared to MVNO m'_n at the high-level stage, i.e., $m_n \succ_n m'_n$. Thus, all members of MVNO m'_n including UE k' are also rejected at iteration τ and cannot be included in the stable group $\mathcal{G}_n$. This implies that no UE k outside the group $\mathcal{G}_n$ can joint it.

Now consider the second condition of Definition 1, i.e., no UE k can leave a stable group. A stable group $\mathcal{G}_n$ is formed by stable pairs (k, m_n) using stable matching following the preference relation in both levels, i.e., $\mu \succ_{m_n} \mu'$ at low-level and $\mu \succ_n \mu'$ at high-level. Thus there exists no matching $\mu'(n)$ and $\mu'(m_i)$ which is better than the current match. Therefore, no UE $k \in \mu(m_n)$ has an incentive to leave the stable group $\mathcal{G}_n$ for any other group $\mathcal{G}'_n$. □

6.4.5 Performance Analysis

To simulate our proposal, we used the MATLAB tool in which we consider the standard parameters of cellular technologies that follow the system guidelines given in [1]. We consider a network with 5 MVNOs that rent slices from N InP-BSs to serve randomly located K UEs inside the coverage area of 1000×1000 m. Each InP owns a band of 1.4 MHz (i.e., six channels or resource blocks). Moreover, the bandwidth W of each channel and weight parameter ω are set to a normalized value of 1. In our simulation, each UE k has a traffic demand (generated randomly) which is uniformly distributed in the range of $d_k = \{1 \sim 3\}$ bps/hz. Note that in this work we do not differentiate between the priority of UEs' traffic demands and assume all users' demand have homogeneous priority. Moreover, we assume that the traffic dynamics do not change during the execution of allocation process. Such an assumption is in line with many existing works [26, 53]. Moreover, we set the prices for MVNOs and InPs that is also uniformly distributed in the range of $\beta_m^M = \{4 \sim 8\}$ and $\beta_n^I = \{2 \sim 4\}$ monetary units/bps/hz, respectively. Furthermore, all results are obtained by averaging over a large number of independent simulation runs (i.e., 500 runs), each of which realizes random traffic demands, pricing, locations of InP-BSs, UEs, and channel power gains.

For comparison purposes, we compare the proposed algorithms with two baseline schemes. First, a fixed sharing scheme (FS), where each MVNO reserves equal number of the channels. This fixed sharing can also be viewed as the case in which there is no wireless virtualization and a comparison of the proposal with FS scheme reflects the benefits achieved by WNV over the traditional cellular networks. Second, a general sharing scheme (GS) in which the MVNOs are not involved and the InP directly performs a single-level matching for the channel allocation, which is in line with some existing works such as [39].

In Fig. 6.9, the average sum-rate versus the network size (i.e., number of UEs) is shown for the different schemes. It is observed that the sum-rate increases with network size, which, however, saturates as the network size becomes sufficiently large. This is due to the limited network bandwidth for each InP (i.e., 1.4 MHz).

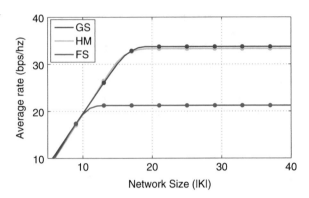

Fig. 6.9 Average sum-rate of HM, GS, and FS schemes

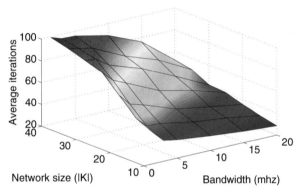

Fig. 6.10 Average iteration vs. network size for varying InP-BS bandwidth

We also observe that the sum-rate obtained by HM and GS schemes results in an indistinguishable performance. Specifically, the HM scheme can achieve up to 97.5% of the average sum-rate obtained by the GS scheme, for a large network size (i.e., $|K| > 20$). Thus, it can be inferred that the HM scheme is close to optimal. Moreover, a performance benefit up to 32% can be achieved when compared to the FS approach for $|K| > 15$. Then, a comparison of average iterations of HM scheme under different network sizes with varying InP-BS bandwidth is shown in Fig. 6.10. The HM scheme achieves convergence under all scenarios in few iterations. However, the iterations increase with the network size because of the increasing number of UE's proposal and accept–reject procedure of HM. Moreover, we also infer that as the InP-BS bandwidth is increased from 1.4 to 20 MHz, the average iterations decrease. This can be explained as bandwidth increases, there are sufficient channels to meet the demands. Therefore, less iterations are required to converge to a stable group as most of the proposals are accepted by the sellers due to large available quota (i.e., channels).

6.5   Matching Game for Non-orthogonal Multiple Access

The explosive growth of data traffic in mobile Internet and the dramatical increase in the number of mobile devices, high spectrum efficiency and massive connectivity in 5G wireless communications will be required. Moreover, the biggest disadvantage of current orthogonal multiple access (OMA) schemes (i.e., the OFDMA scheme) is the number of served users, which will be limited by the number of spectrum resources (i.e., subchannels, resource blocks). NOMA has been considered as a key enabling technique for 5G cellular systems [41, 44], which can alleviate the aforementioned challenge of OMA schemes and boost WNV development by exploiting the spectrum sharing for guaranteeing isolation among multiple MVNOs. In NOMA, by exploiting the channel gain differences, multiple users are multiplexed into the transmission power domain and then non-orthogonally scheduled for transmission over the same spectrum resources. This technique can provide massive device connectivity in comparison to traditional OMA schemes.

In NOMA, users with significantly different channel gains over a resource block are grouped together as shown in (Fig. 6.11). Therefore, our aim is to find a set of users that can be grouped into the same cluster. Note that the number of clusters in a network depends on the network users' channel conditions, i.e., a large number

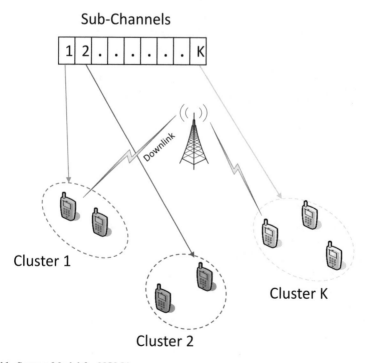

Fig. 6.11 System Model for NOMA

of good channel users (or bad channel users) cannot be grouped together as they would experience strong interference. Thus, our first aim is to classify users based on channel conditions. We classify the network users into two classes of users: A and B corresponding to strong and weak users depending on their respective channel gain. Any number of users' classification can be defined, however, more number of users in a class increases the complexity of the receivers.

6.5.1   System Model

We consider the downlink of a single cell consisting of one MBS. The MBS serves a set of cellular users (CUs) denoted by $\mathcal{M}$ and the number of CUs is M. The MBS works on a system bandwidth which is divided into a set of subchannels denoted $\mathcal{S}$, each of bandwidth B as shown in (Fig. 6.11). In our model, users who are packed or scheduled over non-orthogonal subchannel form a NOMA cluster. Each NOMA cluster operates on a subchannel which is orthogonal to other subchannels allocated to other clusters. Furthermore, the number of users per NOMA cluster can range between 1 and $|\mathcal{M}|$. However, in practical scenarios, the number of users per NOMA cluster is generally set to two in order to reduce the hardware complexity for successive interference cancellation (SIC).

Let $\mathcal{M}_k$ be the set of active CUs grouped into the kth cluster, the maximum MBS transmission power budget is P_T, and the maximum transmission power budget per downlink NOMA cluster is P_t. The power allocated to CU $m \in \mathcal{M}$ is denoted by P_m. The complex coefficient of channel between CU m and the MBS is denoted by $h_m = \chi_m / \mathcal{D}(d_m)$, where χ_m denotes the Rayleigh fading channel gain, $\mathcal{D}(\cdot)$ is the path loss function, and d_m is the geographical distance between CU m and the MBS. Let x_m be the transmitted symbol of CU m. The signal that CU m received from MBS in the kth cluster is then given by

$$y_m^k = h_m \sqrt{P_m} x_m^k + \sum_{m' \neq m | m' \in \mathcal{M}_k} h_m \sqrt{P_{m'}} x_{m'}^k \tag{6.20}$$

where z_m is the additive white Gaussian noise.

Since all CUs belonging to the same cluster can utilize same subchannel allocated to that cluster, the signal of any user causes interference to others. To demodulate the target message, each CU performs SIC after receiving the superposed signals [10]. In general, the users with higher channel gains are allocated low power levels and their signals can be recovered after all users with higher power levels are recovered in the SIC decoding. Similarly, the users with lower channel gains have high power assignment levels and their signals are recovered by treating the users' signals with lower power levels as the noise in the SIC decoding [3, 10].

The optimal order of SIC decoding is in the order of the increasing channel gains normalized by the noise. To be specific, the receiver of CU $m \in \mathcal{M}_k$ can cancel the interference from any other CU $m' \in \mathcal{M}_k$ with channel gain $|h_{m'}|^2 / z_{m'} < |h_m|^2 / z_m$, i.e., CU m first decodes the signal from UE m' then subtracts it and decodes its target signal x_m^k correctly from the received signal y_m^k in kth cluster. We now define a user clustering (grouping) variable β_m^k as follows:

$$\beta_m^k = \begin{cases} 1 & \text{if CU } m \text{ is grouped into cluster } k \\ 0 & \text{otherwise,} \end{cases}$$

Then, the achievable throughput for CU m in downlink NOMA kth cluster can be expressed as

$$R_{m,CU}^k = \log_2 \left(1 + \frac{P_m |h_m|^2}{I_m^k + z_m} \right), \tag{6.21}$$

where I_m^k is the interference that CU $m \in \mathcal{M}_k$ receives due to the other CUs in kth cluster

$$I_m^k = \sum_{m' \in \mathcal{M} | \frac{|h_{m'}|^2}{z_{m'}} > \frac{|h_m|^2}{z_m}} \beta_{m'}^k P_{m'} |h_m|^2. \tag{6.22}$$

6.5.2 Problem Formulation

Our goal is maximize the network throughput by maximizing the number of service users in the network. Thus, we aim to cluster users and present the problem of user clustering as follows:

$$\max_{\beta} \sum_{k \in \mathcal{S}} \left(\sum_{m \in \mathcal{M}} \beta_m^k R_{m,CU}^k \right)$$

s.t.:

$$C_1 : \sum_{k \in \mathcal{S}} \sum_{m \in \mathcal{M}} \beta_m^k P_m \leq P_T, \tag{6.23}$$

$$C_2 : \sum_{k \in \mathcal{S}} \beta_m^k = 1, \ \forall m \in \mathcal{M},$$

$$C_3 : \beta_m^k \in \{0, 1\} \ \forall m \in \mathcal{M},$$

The user clustering problem in (6.23) is still a combinatorial problem, and finding the solution becomes NP-hard, for a large set of users and subchannels in a practical amount of time [27]. Note that problem (6.23) is desired to be solved in a distributed manner. Therefore, we use matching theory to map the problem (6.23) into a matching game which has the ability to solve combinatorial problems. Furthermore, matching theory allows each player (i.e., users and subchannels) to define its individual utilities depending upon its local information. Moreover, for the user clustering problem, we present a distributed matching in which the CUs and subchannels act as players. In this game, our aim is to find the set of CUs that can be grouped into the same subchannels to form a NOMA cluster. The details of this game are discussed in the following sections.

6.5.3 Proposed Solution

In NOMA, users with significantly different channel gains over a subchannel are grouped together to form a NOMA cluster. Therefore, our aim is to find the set of CUs that can be grouped into the same subchannel to form a NOMA cluster.

To classify CUs, it is assumed that the MBS first obtains CSI of all the CUs. Then, it sorts them in a decreasing order. Finally, we can use a pre-defined threshold to classify the users into two classes, i.e., CUs greater than a certain threshold fall into the same class. Users in one class are similar to each other in terms of channel gain and dissimilar to the users belonging to other classes. The similarity between users is based on a measure of the channel gains between them and the MBS.

6.5.3.1 Game Formulation

Once user classification is executed, the next goal here is to perform CUs grouping into clusters. In this game there are two disjoint sets of agents, the set of clusters, $\mathscr{S}$, and the set of CUs, $\mathscr{M}$. Each cluster s has a strict, transitive, and complete preference profile $\mathscr{P}_s$ defined over users, i.e., $2^{\mathscr{M}}$. Note that, in this game, from constraint C_3 in (6.23), it is given that each user can be assigned to a single cluster. However, different users can exist in same cluster, i.e., property of NOMA. Therefore, the preference profile $\mathscr{P}_m$ of CUs is defined over the clusters, i.e., $\mathscr{S}$. Note that other users m'' operating in the same cluster implicitly affect the preference ranking of user m. Therefore, our design corresponds to the *one-to-many matching* given by the tuple $(\mathscr{M}, \mathscr{S}, \succ_{\mathscr{M}}, \succ_{\mathscr{S}})$. Here, $\succ_{\mathscr{M}} \triangleq \{\succ_j\}_{j \in \mathscr{M}}$ and $\succ_{\mathscr{S}} \triangleq \{\succ_k\}_{k \in \mathscr{S}}$ represent the set of the preference relations of the users and clusters, respectively. Formally, we define the matching as follows:

Definition 6 A *matching* β is defined on the set $\mathscr{M} \cup \mathscr{S}$ which satisfies for all $k \in \mathscr{S}$ and $j \in \mathscr{M}$:

1. $|\beta(j)| \leq 1$ and $\beta(j) \in \mathscr{S} \cup \phi$,
2. $|\beta(k)| \leq q_k$ and $\beta(k) \in 2^{\mathscr{M}} \cup \phi$,
3. If $j \in \beta(k)$, then $\beta(j) = k$,
4. If $\beta(j) \in k$ for cluster k, then $\beta(k) = j$,

where q_k denotes the quota of cluster k, $|\beta(.)|$ denotes the cardinality of matching outcome $\beta(.)$. The first condition here represents the constraint C_3 in (6.23). The second condition represents the quota of a subchannel k. In this game, we restrict the value of q_k to 2 CUs to reduce the receiver complexity. Here, $\beta(j) = \phi$ means that j is not matched to any cluster. Similarly, if $\beta(k) = \phi$, then there is no user matched to cluster k.

6.5.3.2 Preference Profiles of Players

In our formulated game, both sides need to rank each other using the preference profile. However, the preference profiles of users here depend on the clusters as well as other users assigned to that cluster. Such interdependence relations are known in matching theory as *externalities*, and have important implications in the design of the proposed solution. Due to these externalities, an agent may continuously change its preference order, in response to the formation of other agents and never reach a final assignment, unless externalities are well-handled.

In order to build the preference profile of users ($\mathscr{P}_j$), each user calculates the achievable data rate for each cluster and then ranks them in descending order. Therefore, the utility of each CU can be defined as follows:

$$U_j(k, \beta) = R^k_{j,CU}, \quad \forall k. \tag{6.24}$$

Thus, for any user j_n, a preference relation $\succ_{j_n}$ is defined over the set of clusters $\mathscr{S}$ such that, for any two clusters $k, k' \in \mathscr{S}, k \neq k'$, and two matchings β and $\beta' \in \mathscr{M} \times \mathscr{S}, k = \beta(j), k' = \beta'(j)$

$$(k, \beta) \succ_j (k', \beta') \Leftrightarrow U_j(k, \beta) > U_j(k', \beta'). \tag{6.25}$$

Similarly, the goal of each cluster is to choose a set of users that can maximize the rate over each cluster k. Therefore, it uses the following utility to create its preference profile ($\mathscr{P}_k$):

$$U_k(\mathscr{A}, \beta) = \sum_{j \in \mathscr{A}} R^k_{j,CU}, \quad \forall \mathscr{A}, \tag{6.26}$$

According to (6.26), each subchannel k chooses a subset of CUs $\mathscr{A}$ that can maximize the achievable rate of a cluster. Moreover, for any cluster k a preference relation $\succ_k$ is defined as follows, for any two subset of CUs $\mathscr{A}, \mathscr{A}' \in \mathscr{M}$, where $\mathscr{A} \neq \mathscr{A}'$, and $\mathscr{A} = \beta(k), \mathscr{A}' = \beta'(k)$:

$$(\mathscr{A}, \beta) \succ_s (\mathscr{A}', \beta') \Leftrightarrow U_k(\mathscr{A}, \beta) > U_k(\mathscr{A}', \beta'). \tag{6.27}$$

Once the matching game and preference profiles of both the agent sides have been defined, we now aim at finding a stable clustering scheme for the proposed game.

However, it is evident from (6.24) and (6.26) that our preferences are a function of the existing matching β and from (6.22), it is clear that users affect each other's performance through interference produced by high SINR users. Therefore, in the next section, we present a novel approach adopted to handle such externalities.

6.5.3.3 Preferences and Externalities

Next, we develop a novel approach to handle externalities in the proposed game and analyze its solution.

In the proposed game if user j is assigned to a cluster k, it will produce interference to the other users using the same cluster k if its gain is higher than those users of clusters. Consequently, an agent may change its preference order with regard to a given cluster k in response to the action of other agents, i.e., user j' which have been assigned to the same cluster k. This may lead to a case in which agents never reach a final clustering.

Therefore, for building the user preference which can also handle the externalities, we propose that the initial network information (i.e., CSI of all UEs) is broadcasted to the CUs by the InPs after collecting it from each individual CUs. Through this information, each user can find the set of CUs that have a higher gain with MBS. Note that, in NOMA, we aim to cluster CUs that have significant difference in channel gains. Then, we assume that CUs only care about other CUs that fall in the same class. Moreover, each CU would have a different set of CUs. We name this set as an externality set for user j that has a set of conflicting users and represent it by $\mathcal{C}_j$ as follows:

$$\mathcal{C}_j = \left\{ j' \in \mathcal{M} : \frac{|h_{j'}|^2}{z_{j'}} > \frac{|h_j|^2}{z_j}, \, j, j' \in \mathcal{B} \right\}, \tag{6.28}$$

where $\mathcal{B}$ represents the set of CUs that fall in the same class. From (6.28), we select the CUs that belong to the same class of user j and have a higher gain compared to user j. The main idea is to restrict the users that belong to the same class to be grouped into same cluster.

6.5.3.4 Proposed Algorithm

In order to find a stable clustering scheme, we need to first define the blocking pair for our game. Note that in the formulated game there is an additional challenge of externalities. Thus, traditional solution designed for one to many games based on Gale-Shapley does not apply over our game. Therefore, first, we design the blocking pair for the formulated game with externalities followed by a stable algorithm. The blocking pair for the formulated game is defined as follows:

Definition 7 A matching β is said to be *stable* if there exists no *blocking pair* (j, k) such that $j \succ_k \beta(k)$, $k \succ_j \beta(j)$, and $\beta(k) \notin \mathcal{C}_j$.

Definition 7 is based on the following intuition. Whenever a user j prefers a cluster s over its assigned cluster $\beta(j)$ that does not contain a conflicting user (i.e., $\beta(k) \notin \mathcal{C}_j$), and cluster k is also willing to admit j (i.e., $k \succ_k \beta(k)$) by rejecting some accepted users in $\beta(k)$ which are ranked lower than j, then j and k can deviate from their assigned matching to form a blocking pair. A matching is stable only if there exists no blocking pair. Moreover, to achieve stability, a sufficient condition is that formation of any new agent pair does not undermine the stability of existing matched pairs. By employing such a condition, the preference profile of currently matched users on a cluster will remain unaltered even after this new pair formation. Stability in our solution ensures that after clustering, no matched pair (user-cluster) in the network would benefit from replacing their assigned cluster with a new better cluster, and vice versa. This property is important to ensure the stable matching for one to many matching problems with externalities.

Next, we present a novel and stable user clustering algorithm. In this algorithm, the MBS first decides the proposing order based on the set of available classes. The intuition behind this assumption comes from the fact that in NOMA we would like to restrict users from same class to be in the same cluster. Thus by allowing a sequential proposing manner in terms of classes will allow same class users to compete with each other. Therefore, in our algorithm, we assume proposal starting by the strongest class A to the weakest class B. Note that by allowing this proposing order, we can guarantee that no matched user from a higher class can be affected by lower class users. The algorithm starts by using the local information to build the preference profiles (lines 1–3). At each iteration t, each user j that belongs to a specific class first calculates its utility and re-ranks all the clusters based on the previous matching $\beta(k)^{(t-1)}$ (line 4).

Then, each user j proposes to the most preferred cluster k (line 6). On receiving the proposals, each cluster k first investigates if there exists a conflicting user j' in $\beta(k)$ that can result in either of the two cases. The first case is that there exists a conflicting user, i.e., $\mathscr{C}_k$ is non-empty (line 8). In this case, k removes all lower ranked users j' compared to j from its current matching (lines 9–12) and rechecks the conflict set (line 13). If still conflict set is non-empty, j is also rejected along with other removed users and is considered as the least preferred user j_{lp} (line 14) otherwise it is accepted (lines 15–16). The second case is the conflict set is empty. In this case, the quota of cluster k (q_k[4]) is first checked, if enough quota exists to accommodate j, then user j is accepted by the cluster k, otherwise user j is rejected by the cluster k and considered as least preferred (lines 17–21). Finally, the least preferred user, i.e., j_{lp} and all users ranked lower than j_{lp} are removed from $\mathscr{P}_k^{(t)}$, and similarly these users also remove k from their respective $\mathscr{P}_j^{(t)}$ (lines 22–24). Once all users of a class have either been accepted or rejected by all the clusters, the next class starts the proposal process. Note that here the matching for a specific class terminates when the results of two consecutive iterations t remain unchanged (line 25). *With this process, we guarantee that any less preferred user will not be accepted by that cluster even if it has sufficient quota to do so, which is crucial for the matching stability of our design.* This process is repeated until the matching converges.

Theorem 3 *Algorithm 3 converges to a stable allocation.*

---

[4]The quota q_k is set to two users per clusters as we assume two classes.

Algorithm 3 Matching-based User Clustering Algorithm

1: **input:** $\mathscr{P}_j^{(t)}$, $\mathscr{P}_k^{(t)}$, $\mathscr{C}_j$, $\forall k, j$.

2: **initialize:** $t = 0$, $\beta^{(1)} \triangleq \{\beta(j)^{(1)}, \beta(k)^{(1)}\}_{j\in\mathcal{M}, k\in\mathcal{S}} = \emptyset$, $\mathscr{J}_k^{(1)} = \emptyset$, $C_s^{(1)} = \emptyset$, $q_k^{(1)} = |\mathscr{U}|$, $\forall k, j$.

3: **repeat**

4:     $t \leftarrow t + 1$.

5:     Update $\forall j$, $\mathscr{P}_j^{(t)}$ for given $\beta(s)^{(t-1)}$.

6:     $\forall j \in$ same class with k as its most preferred in $\mathscr{P}_j^{(t)}$.

7:     **while** $j \notin \beta(k)^{(t)}$ and $\mathscr{P}_j^{(t)} \neq \emptyset$ **do**

8:         **if** $C_k^{(t)} = \{j' \in \beta(k)^{(t)} \cup C_j\} \neq \emptyset$ **then**

9:             $\mathscr{X'}_k^{(t)} = \{j' \in \beta(k)^{(t)}, k' \in \mathscr{C}_j | j \succ_k j'\}$.

10:            $j_{lp} \leftarrow$ the least preferred $j' \in \mathscr{X'}_k^{(t)}$.

11:            **for** $j_{lp} \in \mathscr{X'}_k^{(t)}$ **do**

12:                $\beta(k)^{(t)} \leftarrow \beta(k)^{(t)} \setminus j_{lp}$, $q_k^{(t)} \leftarrow q_k^{(t)} + 1$.

13:            **if** $C_k^{(t)} = \{j' \in \beta(k)^{(t)} \cup C_j\} \neq \emptyset$ **then**

14:                $j_{lp} \leftarrow j$.

15:            **else**

16:                $\beta(k)^{(t)} \leftarrow \beta(s)^{(t)} \cup j$, $q_k^{(t)} \leftarrow q_k^{(t)} - 1$.

17:         **else**

18:            **if** Check $q_k^{(t)} > 0$ **then**

19:                $\beta(k)^{(t)} \leftarrow \beta(k)^{(t)} \cup j$, $q_k^{(t)} \leftarrow q_k^{(t)} - 1$.

20:            **else**

21:                $j_{lp} \leftarrow j$.

22:         $\mathscr{J}_k^{(t)} = \{j \in \mathscr{X'}_k^{(t)} | j_{lp} \succ_k j\} \cup \{k_{lp}\}$.

23:         **for** $j \in \mathscr{J}_r^{(t)}$ **do**

24:            $\mathscr{P}_j^{(t)} \leftarrow \mathscr{P}_j^{(t)} \setminus k \mathscr{P}_k^{(t)} \leftarrow \mathscr{P}_k^{(t)} \setminus j$.

25:     **Check:** $\beta^{(t-1)} = \beta^{(t)}$.

26: **until** $\forall$ classes, i.e., A, B.

Proof We prove this theorem by contradiction. Assume that Algorithm 3 produces a matching β with a blocking pair (j, k) by Definition 7. Since $k \succ_j \beta(j)$, j must have proposed to k and has been rejected due to a more preferred conflicting user j' on cluster k (lines 13–14). Thus, in this case (j, k) cannot form a blocking pair as $j' \succ_k j$, a contradiction. Moreover, when j was rejected, then any lower ranked user j' was rejected either before j (lines 9–12) or was made unable to propose because k is removed from j' preference list (lines 22–24). Thus, any lower ranked j' cannot be matched by k, i.e., $j' \notin \beta(r)$, a contradiction. $\qquad\square$

6.5.4 Performance Analysis

We consider a downlink system in which the BS is assumed to be deployed at a fixed location, and we randomly deploy C cellular users following a homogeneous Poisson point process (PPP). We assume the system bandwidth to be 3 MHz. Note

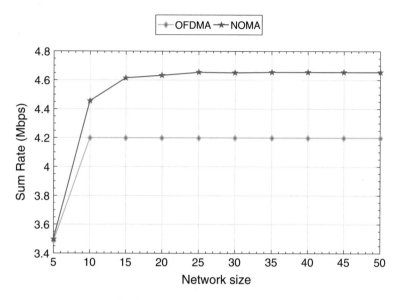

Fig. 6.12   Average sum-rate of NOMA and OFDMA schemes

that the methodologies developed in this work can also be applied to any value of system bandwidth. The motivation for our choice (i.e., 3 MHz) is to analyze the performance under dense environment with peak network traffic and for the sake of simulation simplicity. Moreover, the wireless parameters are chosen according to the system model guidelines in [1]. Moreover, we compare the performance of our proposed scheme with the traditional OFDMA scheme. In the OFDMA scheme, users are assigned with an orthogonal number of RBs, i.e., no interference between the users. Note that all statistical results are averaged over 500 runs of random locations of cellular users, and RB gains.

In Fig. 6.12, we illustrate the total network sum rate vs. the number of users under two schemes, NOMA and OFDMA. We evaluate the total throughput by obtaining the average total sum-rate over different number of users. It can be seen that the total throughputs of NOMA increase when the number of users increases until it becomes saturated, i.e., when the number of users in the network increases significantly. It can be observed that when the number of users is larger than 15, the total throughput continues to increase due to the multiuser diversity gain, but grows at a slower speed and becomes saturated when the number of users is sufficiently large. Moreover, we can see that the performance of proposed scheme outperforms the traditional OFDMA scheme significantly. This signifies the importance of NOMA in bringing 5G networks into fruition.

In Fig. 6.13, we compare the number of admitted users in the system for the proposed NOMA and OFDMA schemes by varying the number of network users. It can be seen that only 15 users (i.e., 3 MHz bandwidth) can be accepted using the OFDMA scheme, whereas in the NOMA the number of admitted users is

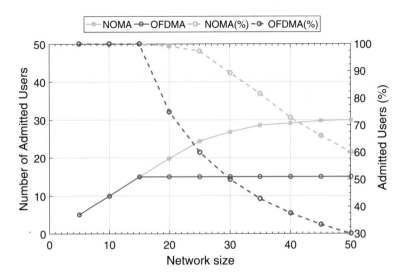

Fig. 6.13 Average admitted users vs. network size

significantly higher, i.e., double of the OFDMA scheme. Thus, we can infer that NOMA will play a crucial role in enhancing the number of admitted users in the 5G networks.

6.6 Conclusions

This chapter provides a comprehensive overview of matching theory and its role in bringing novel 5G networking paradigms into fruition. Specifically, we have provided theoretical analysis for three 5G networking paradigms such as dense heterogeneous networks, wireless network virtualization, and non-orthogonal multiple access. Moreover, we have also provided the numerical analysis for all these paradigms. Numerical results reveal that matching theory can significantly enhance the performance of the 5G networks in terms of average sum-rate and average number of admitted users. Furthermore, the distributed control of matching games will play a very crucial role in realizing many novel applications for future wireless networks.

Acknowledgement This work was supported by the National Research Foundation of Korea(NRF) grant funded by the Korea government(MSIT) (NRF-2017R1A2A2A05000995).

References

1. 3GPP: Evolved universal terrestrial radio access (E-UTRA): physical layer procedures, Release 11. Technical Report TS 36.213 (2012)
2. Abdelnasser, A., Hossain, E., Kim, D.I.: Tier-aware resource allocation in OFDMA macrocell-small cell networks. IEEE Trans. Commun. **63**(3), 695–710 (2015)
3. Ali, M.S., Tabassum, H., Hossain, E.: Dynamic user clustering and power allocation for uplink and downlink non-orthogonal multiple access (NOMA) systems. IEEE Access **4**, 6325–6343 (2016)
4. Alliance, N.G.M.N.: 5G white paper. Next generation mobile networks, White Paper (2015)
5. Andrews, J.G., Buzzi, S., Choi, W., Hanly, S.V., Lozano, A., Soong, A.C., Zhang, J.C.: What will 5G be? IEEE J. Sel. Areas Commun. **32**(6), 1065–1082 (2014)
6. Boyd, S., Vandenberghe, L.: Convex Optimization. Cambridge University Press, Cambridge (2004)
7. Chen, Y., Zhang, J., Wu, K., Zhang, Q.: TAMES: a truthful double auction for multi-demand heterogeneous spectrums. IEEE Trans. Parallel Distrib. Syst. **25**(11), 3012–3024 (2014)
8. Cisco: Cisco Visual Networking Index: Global Mobile Data Traffic Forecast Update, 2016–2021 White Paper (2017). Available http://www.cisco.com/c/en/us/solutions/collateral/service-provider/visual-networking-index-vni/mobile-white-paper-c11-520862.html. Cited 28 Mar 2017
9. Demestichas, P., Georgakopoulos, A., Karvounas, D., Tsagkaris, K., Stavroulaki, V., Lu, J., Xiong, C., Yao, J.: 5G on the horizon: key challenges for the radio-access network. IEEE Veh. Technol. Mag. **8**(3), 47–53 (2013)
10. Di, B., Song, L., Li, Y.: Sub-channel assignment, power allocation, and user scheduling for non-orthogonal multiple access networks. IEEE Trans. Wirel. Commun. **15**(11), 7686–7698 (2016)
11. Echenique, F., Oviedo, J.: A theory of stability in many-to-many matching. Theor. Econ. **1**, 233–273 (2006)
12. Elshaer, H., Boccardi, F., Dohler, M., Irmer, R.: Downlink and uplink decoupling: a disruptive architectural design for 5G networks. In: IEEE Global Communications Conference (GLOBECOM), Austin (2014)
13. Fu, F., Kozat, U.C.: Stochastic game for wireless network virtuazlization. IEEE/ACM Trans. Networking **21**(1), 84–97 (2013)
14. Gale, D., Shapley, L.S.: College admissions and the stability of marriage. Am. Math. Mon. **69**(1), 9–15 (1962)
15. Gu, Y., Saad, W., Bennis, M., Debbah, M., Han, Z.: Matching theory for future wireless networks: fundamentals and applications. IEEE Commun. Mag. **53**(5), 52–59 (2015)
16. Hamidouche, K., Saad, W., Debbah, M.: Multi-game framework for harmonized LTE-U and WiFi coexistence over unlicensed bands. IEEE Wirel. Commun. Mag. **23**(6), 62–69 (2016)
17. Han, Z., Gu, Y., Saad, W.: Matching Theory for Wireless Networks. Springer, Cham (2017)
18. Hausken, K., Cressman, R.: Formalization of multi-level games. Int. Game Theory Rev. **6**(02), 195–221 (2004)
19. Ho, T.M., Tran, N.H., Kazmi, S.A., Hong, C.S.: Dynamic pricing for resource allocation in wireless network virtualization: a Stackelberg game approach. In: The International Conference on Information Networking (ICOIN), Da Nang (2017)
20. Ho, T.M., Tran, N.H., Kazmi, S.A., Han, Z., Hong, C.S.: Wireless network virtualization with non-orthogonal multiple access. In: IEEE/IFIP Network Operations and Management Symposium, Taipei (2018)
21. Hong, C.S., Kazmi, S.A., Moon, S., Van Mui, N.: SDN based wireless heterogeneous network management. In: AETA 2015: Recent Advances in Electrical Engineering and Related Sciences. Springer, Cham (2016)
22. Hossain, E., Hasan, M.: 5G cellular: key enabling technologies and research challenges. IEEE Instrum. Meas. Mag. **18**(3), 11–21 (2015)

23. Kamel, M.I., Le, L.B., Girard, A.: LTE wireless network virtualization: dynamic slicing via flexible scheduling. In: Proceedings of the IEEE 80th Vehicular Technology Conference (VTC), Vancouver (2014)
24. Kazmi, S.A., Hong, C.S.: A matching game approach for resource allocation in wireless network virtualization. In: The International Conference on Ubiquitous Information Management and Communication (IMCOM), Beppu (2017)
25. Kazmi, S.A., Tran, N.H., Ho, T.M., Oo, T.Z., LeAnh, T., Moon, S., Hong, C.S.: Resource management in dense heterogeneous networks. In: 17th Asia-Pacific Network Operations and Management Symposium, APNOMS, Busan (2015)
26. Kazmi, S.A., Tran, N.H., Saad, W., Le, L.B., Ho, T.M., Hong, C.S.: Optimized resource management in heterogeneous wireless networks. IEEE Commun. Lett. **20**(7), 1397–1400 (2016)
27. Kazmi, S.A., Tran, N.H., Saad, W., Han, Z., Ho, T.M., Oo, T.Z., Hong, C.S.: Mode selection and resource allocation in device-to-device communications: a matching game approach. IEEE Trans. Mob. Comput. **16**(11), 3126–3141 (2017)
28. Kazmi, S.A., Tran, N.H., Ho, T.M., Hong, C.S.: Hierarchical matching game for service selection and resource purchasing in wireless network virtualization. IEEE Commun. Lett. **22**(1), 121–124 (2018)
29. Kelly, F.P., Maulloo, A.K., Tan, D.K.: Rate control for communication networks: shadow prices, proportional fairness and stability. J. Oper. Res. Soc. **49**(3), 237–252 (1998)
30. Lei, L., Yuan, D., Ho, C.K., Sun, S.: Power and channel allocation for non-orthogonal multiple access in 5G systems: tractability and computation. IEEE Trans. Wirel. Commun. **15**(12), 8580–8594 (2016)
31. Liang, C., Yu, F.R.: Distributed resource allocation in virtualized wireless cellular networks based on ADMM. In: IEEE Conference on Computer Communications Workshops (INFOCOM WKSHPS), Hong Kong (2015)
32. Liang, C., Yu, F.R.: Wireless network virtualization: a survey, some research issues and challenges. IEEE Commun. Surv. Tutorials **17**(1), 358–380 (2015)
33. Liu, B., Tian, H.: A bankruptcy game-based resource allocation approach among virtual mobile operators. IEEE Commun. Lett. **17**(7), 1420–1423 (2013)
34. Lopez-Perez, D., Guvenc, I., De la Roche, G., Kountouris, M., Quek, T.Q., Zhang, J.: Enhanced intercell interference coordination challenges in heterogeneous networks. IEEE Wirel. Commun. **18**(3), 22–30 (2011)
35. Maghsudi, S., Stanczak, S.: Joint channel allocation and power control for underlay D2D transmission. In: IEEE International Conference on Communications (ICC), London (2015)
36. Manlove, D.F.: Algorithmics of Matching Under Preferences. World Scientific, Singapore (2013)
37. Osseiran, A., Boccardi, F., Braun, V., Kusume, K., Marsch, P., et al.: Scenarios for 5G mobile and wireless communications: the vision of the METIS project. IEEE Commun. Mag. **52**(5), 26–35 (2014)
38. Panwar, N., Sharma, S., Singh, A.K.: A survey on 5G: the next generation of mobile communication. Phys. Commun. **18**, 64–84 (2015)
39. Parsaeefard, S., Dawadi, R., Derakhshani, M., Le-Ngoc, T.: Joint user-association and resource-allocation in virtualized wireless networks. IEEE Access **4**, 2738–2750 (2016)
40. Roth, A.E.: Deferred acceptance algorithms: history, theory, practice, and open questions. Int. J. Game Theory **36**(3–4), 537–569 (2008)
41. Saito, Y., Kishiyama, Y., Benjebbour, A., Nakamura, T., Li, A., Higuchi, K.: Non-orthogonal multiple access (NOMA) for cellular future radio access. In: Proceedings of the IEEE Vehicular Technology Conference (VTC), Dresden (2013)
42. Son, K., Lee, S., Yi, Y., Chong, S.: REFIM: a practical interference management in heterogeneous wireless access networks. IEEE J. Sel. Areas Commun. **29**(6), 1260–1272 (2011)
43. Song, L., Niyato, D., Han, Z., Hossain, E.: Game-theoretic resource allocation methods for device-to-device communication. IEEE Wirel. Commun. **21**(3), 136–144 (2014)

44. Song, L., Li, Y., Ding, Z., Poor, H.V.: Resource management in non-orthogonal multiple access networks for 5G and beyond. IEEE Netw. **31**(4), 8–14 (2017)
45. Ullah, S., Thar, K., Hong, C.S.: Management of scalable video streaming in information centric networking. Multimed. Tools Appl. (2017). https://doi.org/10.1007/s11042-016-4008-8
46. Van De Belt, J., Ahmadi, H., Doyle, L.E.: A dynamic embedding algorithm for wireless network virtualization. In: Proceedings of the IEEE 80th Vehicular Technology Conference (VTC), Vancouver (2014)
47. Venturino, L., Prasad, N., Wang, X.: Coordinated scheduling and power allocation in downlink multicell OFDMA networks. IEEE Trans. Veh. Technol. **58**(6), 2835–2848 (2009)
48. Wang, C.X., Haider, F., Gao, X., You, X.H., Yang, Y., Yuan, D., et al.: Cellular architecture and key technologies for 5G wireless communication networks. IEEE Commun. Mag. **52**(2), 122–130 (2014)
49. Wei, G., Vasilakos, A.V., Zheng, Y., Xiong, N.: A game-theoretic method of fair resource allocation for cloud computing services. J. Supercomput. **54**(2), 252–269 (2010)
50. Xu, H., Li, B.: Anchor: a versatile and efficient framework for resource management in the cloud. IEEE Trans. Parallel Distrib. Syst. **24**(6), 1066–1076 (2013)
51. Yuan, P., Xiao, Y., Bi, G., Zhang, L.: Towards cooperation by carrier aggregation in heterogeneous networks: a hierarchical game approach. IEEE Trans. Veh. Technol. **66**(2), 1670–1683 (2017)
52. Zhang, H., Xiao, Y., Bu, S., Niyato, D., Yu, F.R., & Han, Z.: Computing resource allocation in three-tier IoT fog networks: a joint optimization approach combining Stackelberg game and matching. IEEE Internet Things J. **4**(5), 1204–1215 (2017)
53. Zhu, K., Hossain, E.: Virtualization of 5G cellular networks as a hierarchical combinatorial auction. IEEE Trans. Mob. Comput. **15**(10), 2640–2654 (2016)

Chapter 7
Enhanced Design of Stochastic Defense System with Mixed Game Strategies

Song-Kyoo Kim

7.1 Introduction

Game theory is a mathematical model that has been applied to various strategic situations of conflict and cooperation between system operations. A mixed strategy is a probability distribution over all possible pure strategies. Consider two players, player 1 and player 2, who plays against each other. It may be that player 1 is uncertain about the behavior of his opponent (i.e., player 2), which implies that the behavior of the opponent is random rather than fixed. After a player has determined a mixed strategy at the beginning of the game by randomizing, that player might pick one of those pure strategies and then adhere to it. The fluctuation theory is determining the behavior of one- and two-dimensional marked point processes on some fixed level. For example, Dhsalalow [1] developed a joint transformation of the first exceed level, first passage time, and the index of the point process. Basically, the observation process checks whether the system crashed, and indicates when is the moment to make the decision for a preliminary operation prepared before the system crashes.

In the past decade, we have observed intense cyber attacks, some of which have hindered proper operations of the network systems. The denial of services (DoS) is one of the common cyber attacks which targeted for specific network systems including servers, media gateways, routers even WiFi hotspots. In a denial-of-service (DoS) attack, an attacker attempts to prevent legitimate users from accessing information or services. It makes a network system to deny its service operations by flooding data to overflow the system capacity [7]. Thus, it stands to reason to offer defense operations that could avoid the crash of a network system by a DoS attack.

S.-K. Kim (✉)
Abu Dhabi School of Management, Abu Dhabi, United Arab Emirates
e-mail: s.kim@adsm.ac.ae

© Springer Nature Switzerland AG 2019
J. B. Song et al. (eds.), *Game Theory for Networking Applications*,
EAI/Springer Innovations in Communication and Computing,
https://doi.org/10.1007/978-3-319-93058-9_7

This article lays out a foundation for predicting the time and caliber of potentially destructive shutdown to a system by means of operational calculus. In this study, one player (player 1) makes the decision at the present time and mainly deals with decision making while the other player (player 2) deals with the future moment, which is uncontrollable. The brief for player 1 is given by the probability distribution over strategies of the opponent, mostly obtained by analytically solving the special process rather than from statistical data. Solving the probability distribution of the opponent by using the basic fluctuation theory [1, 5] is another core contribution of this research.

7.2 Mixed Strategy Under Stochastic Time Series

7.2.1 Preliminaries

The moment of the preliminary operation is one step before crashing the system being the same as the time for decision making. The second part is finding the probability of the system being crashed when the next observation moment comes. Let us consider a two-person mixed strategy game, and the player 1 is the person who has two strategies at the observation moment, one step before crashing the system. Player 1 has the following strategies: (1) DoNothing—doing nothing, which implicates that the system is running as usual, and (2) Action—taking the preliminary action for preparing the damage (or crash). In the view of player 2, the system could either crash or be ongoing at the expected observation moment right after the system crash, and the strategies for player 2 would be either "Running" or "Crashed." Let us assume that the cost for the defense operation is α and, if the crash occurred, it would take β amount of this cost. As such, it is rational to assume that the cost for the defense operation will be smaller than cost of encountering a shutdown and recovery; this gives us the normal form games:

. Players: $N = \{1, 2\}$,
. Strategy sets: $S_1 = \{\text{"DoNothing", "Action"}\}$,
                  $S_2 = \{\text{"Running", "Crashed"}\}$,

Based on the above conditions, the general payoff matrix could be composed as follows:

and the brief of Player 1 depends on the probability of the damage which denoted by Γ^*. It is noted that the operation payoffs could be different, depending on the actual payoff status of the system operations, and the payoff matrix (Table 7.1) should be composed accordingly.

Table 7.1 Payoff matrix

	Running	Crashed
DoNothing	0	$-\beta$
Action	$-\alpha$	$-\alpha$

Let C_k be the number of quantity that indicates the level of the system damage when the number passes the limit (i.e., first exceed level) and D_k be the duration during the kth observation period $[\tau_{k-1}, \tau_k)$. The player will make the decision at some τ_{k-1} if $A_{k-1} = C_0 + C_1 + \cdots + C_{k-1} < S$, while $A_k \geq S$. A value of k for which the first excess occurs at τ_k is called the termination index which denoted by ν [1]. It is the moment when the crash (or disaster) is observed (i.e., first observation moment after crashing) which is called the first passage time (denoted by τ_ν). In addition, the moment one step before the first passage time is considered as the decision point for the preliminary action which denoted by $\tau_{\nu-1}$. The probability of the observation of right after the system damage Γ^* could be determined as follows:

$$\Gamma^*_{\text{Idle}} = \mathbb{P}\left\{\tau_\nu \geq \bar{\tau}\right\}, \; \bar{\tau} := \mathbb{E}[\tau_\nu], \tag{7.1}$$

and, automatically, the probability of sustaining the system is $1 - \Gamma^*$. Based on the payoff matrix and the crashing probability, Nash equilibrium of the mixed strategy is as follows:

$$0 \cdot \left(1 - \Gamma^*\right) - \beta \cdot \Gamma^* = -\alpha \cdot \left(1 - \Gamma^*\right) - \alpha \cdot \Gamma^*, \tag{7.2}$$

$$\Gamma^*_{\text{payoff}} = \frac{\alpha}{\beta}.$$

Therefore, we can clearly note the analogy between the two players. The best strategy for the player 1 at moment $(\tau_{\nu-1})$ depends on the chance of crashing the system and the payoff values. The criteria for choosing the best strategy is as follows from (7.2):

$$s_1^* = \begin{cases} \text{``DoNothing''}, & \Gamma^* \leq \frac{\alpha}{\beta}, \\ \text{``Action''}, & \Gamma^* > \frac{\alpha}{\beta}. \end{cases} \tag{7.3}$$

Player 1 would have the chance to make the strategic decision whether to take the preliminary operation or not at one step before exceeding the system capacity (i.e., $\tau_{\nu-1}$), and the decision will be taken based on the above criteria. Basically, the defense operation will be taken at the moment $\tau_{\nu-1}$ before the crash occurs ($A_k \geq S$ and S is the limit considered as "crash of the system"). The more rigorous description of this process is rendered by means of a generic marked point process:

$$(A, \tau) = \sum_{k \geq 0} C_k \cdot \varepsilon_{\tau_k}, \tag{7.4}$$

where ε_a is the point mass with the position (τ) dependent marking. The two-variated sequences of random vectors $(C_k, D_k = \tau_k - \tau_{k-1})$ are iid. The joint transformation

$$\mu\,(z,\theta) = \mathbb{E}\left[z^{C_k}e^{-\theta D_k}\right] \tag{7.5}$$

of the random vectors (C_k, D_k) could be found. The initial observation moment after the action of exiting status may have different but similar from (7.5):

$$\mu_0\,(z,\theta) = \mathbb{E}\left[z^{C_0}e^{-\theta \tau_0}\right]. \tag{7.6}$$

To describe the behavior of the process at the first passage time, the termination index is:

$$\nu = \min\{k : A_k \geq S\}. \tag{7.7}$$

The joint functional of the first exceed model is as follows:

$$L_S\,(\xi, z, \theta, \vartheta) = \mathbb{E}\left[\xi^\nu z^{A_\nu}e^{-\theta \tau_\nu}e^{-\vartheta \tau_{\nu-1}}\right], \tag{7.8}$$

where S indicates the observation moment that is the first moment after crashing the system and (7.8) is solved by using the first exceed level process [1]. The operator D_p is defined as

$$D_p\,f\,(p) = (1-z)\sum_{p\geq 0} f\,(p)\,z^p,\ \|z\| < 1, \tag{7.9}$$

where $\{f\,(p)\}$ is a sequence, with inverse

$$\mathfrak{D}_x^k\,(\bullet) = \begin{cases} \frac{1}{k!}\lim_{x\to 0}\frac{\partial^k}{\partial x^k}\frac{1}{1-x}\,(\bullet), & k \geq 0, \\ 0, & k < 0. \end{cases} \tag{7.9a}$$

The functional $\mathfrak{D}$ from (7.9a) is defined on the space of all analytic functions at 0 and has the following properties:

(i) $\mathfrak{D}_x^k$ is a continuous and linear functional with fixed points at constant functions,
(ii)

$$\mathfrak{D}_x^k\left(x^m f\,(x)\right) = \begin{cases} \mathfrak{D}_x^{k-m}\left(f(x)\right), & k \geq m, \\ 0, & k < m, \end{cases} \tag{7.9b}$$

(iii)

$$\mathfrak{D}_x^k\sum_{m=0}^{\infty} a_m x^m = \sum_{m=0}^{k} a_m. \tag{7.9c}$$

Since the observation is a marked process, the probability of crash at the first exceed observation time is equivalent with the termination index of the first exceed observation:

$$\mathbb{P}\left\{\tau_\nu \leq \bar{\tau}\right\} = \mathbb{P}\left\{\nu \geq \bar{\nu}\right\}, \ \bar{\nu} := \mathbb{E}\left[\nu\right]. \tag{7.10}$$

It is notable that finding the probability of the index is easier than the probability of the first exceed level period. This probability is used for finding the probability of crashing, Γ^*.

7.2.2 The Moment of the Decision for Preliminary Operations

This section utilizes analysis of the random vector and the explicit probability distribution function for the first exceed observation moment after the system crashes. The main result is included in the following theorems:

Theorem 1 *The functional $L_S(\xi, z, \theta, \vartheta)$ of (8) satisfies the following expression:*

$$L_S(\xi, z, \theta, \vartheta) = \mathfrak{D}_w^S \frac{\mu_0(wz, \theta + \vartheta)}{\mu(wz, \theta + \vartheta)} \cdot \frac{\mu(z, \theta) - \mu(wz, \theta)}{1 - \xi\mu(wz, \theta + \vartheta)}. \tag{7.11}$$

Proof We find the explicit formula of the joint function $L_S(\xi, z, \theta, \vartheta)$. The joint functional (7.11) is as follows:

$$L(\xi, z, \theta, \vartheta) = \sum_{n=0}^{\infty} \xi^n \mathbb{E}\left[1_{\{\nu_p=n\}} z^{A_n} e^{-\theta\tau_n} e^{-\vartheta\tau_{n-1}}\right],$$

and

$$\Psi(\xi, z, \theta, \vartheta; w) = (1 - w)\sum_{p\geq 0} L(\xi, z, \theta, \vartheta)$$

$$= \sum_{n=0}^{\infty} \xi^n \mathbb{E}\left[(wz)^{A_{n-1}} e^{-(\theta+\vartheta)\tau_{n-1}}\right] \mu(z, \theta)$$

$$- \sum_{n=0}^{\infty} \xi^n \mathbb{E}\left[(wz)^{A_{n-1}} e^{-(\theta+\vartheta)\tau_{n-1}}\right] \mu(wz, \theta),$$

and, from [5],

$$\mathbb{E}\left[(wz)^{A_{n-1}} e^{-(\theta+\vartheta)\tau_{n-1}}\right] = \mu_0(wz, \theta + \vartheta)\left[\mu(wz, \theta + \vartheta)\right]^{n-1}.$$

Finally, we have

$$\Psi\left(\xi, z, \theta, \vartheta; w\right) = \frac{\mu_0\left(wz, \theta + \vartheta\right)}{\mu\left(wz, \theta + \vartheta\right)} \cdot \frac{\mu\left(z, \theta\right) - \mu\left(wz, \theta\right)}{1 - \xi \mu\left(wz, \theta + \vartheta\right)}. \tag{7.12}$$

From (7.9)–(7.9c), the joint transform of the vector satisfies the formula

$$L_S\left(\xi, z, \theta, \vartheta\right) = \mathfrak{D}_w^S \Psi\left(\xi, z, \theta, \vartheta; w\right). \tag{7.13}$$

From (7.12)–(7.13),

$$L_S\left(\xi, z, \theta, \vartheta\right) = \mathfrak{D}_w^S \left[\frac{\mu_0\left(wz, \theta + \vartheta\right)}{\mu\left(wz, \theta + \vartheta\right)} \cdot \frac{\mu\left(z, \theta\right) - \mu\left(wz, \theta\right)}{1 - \xi \mu\left(wz, \theta + \vartheta\right)}\right]. \tag{7.14}$$

$\square$

Theorem 2 *The probability distribution of the termination index v denoted by $\gamma_k := \mathbb{P}\left[v = k\right]$ satisfies the following expression:*

$$\gamma_k = \mathfrak{D}_w^S \left(\mu_0\left(w, 0\right) \left[1 - \mu\left(w, 0\right)\right] \mu\left(w, 0\right)^{k-1}\right), \quad k = 0, 1, \ldots, \tag{7.15}$$

Proof From (7.8), the probability generating function $g\left(\xi\right)$ of v could be found as follows:

$$g\left(\xi\right) := \mathbb{E}\left[\xi^v\right] = L_S\left(\xi, 1, 0, 0\right), \tag{7.16}$$

and

$$g\left(\xi\right) = \sum_{k \geq 0} \gamma_k \cdot \xi^k = \sum_{l=0}^{\infty} \left[\mathfrak{D}_w^S \left[\frac{\mu_0\left(w, 0\right) \left[1 - \mu\left(w, 0\right)\right]}{\mu\left(w, 0\right)}\right] \mu\left(w, 0\right)^l\right] \xi^l. \tag{7.17}$$

Therefore,

$$\gamma_k = \mathfrak{D}_w^S \left(\mu_0\left(w, 0\right) \left[1 - \mu\left(w, 0\right)\right] \mu\left(w, 0\right)^{k-1}\right). \tag{7.18}$$

$\square$

The decision parameters such as the first exceed observation index v and the period between starting and hitting the crash level are as follows:

$$\bar{v} = \mathbb{E}\left[v\right] = \mathfrak{D}_w^S \frac{\mu_0\left(w, 0\right)}{1 - \mu\left(w, 0\right)}, \tag{7.19}$$

$$\bar{\tau}_{v-1} = \mathbb{E}\left[\tau_{v-1}\right] = \mathfrak{D}_w^S \frac{\mu'\left(w, 0\right) \left(1 - \mu\left(w, 0\right)\right) \left(1 - 2\mu\left(w, 0\right)\right)}{\left(\mu\left(w, 0\right) - \mu^2\left(w, 0\right)\right)^2}. \tag{7.20}$$

Table 7.2 Detailed description of the parameters

Parameter	Description
$\bar{\tau}_{\nu-1}$	$(= \mathbb{E}[\tau_{\nu-1}])$ The moment when Player 1 makes the decision
$\bar{\nu}$	$(= \mathbb{E}[\nu])$ The average number of observation until the system is crushed
Γ^*	$(= \mathbb{P}\{\nu \geq \bar{\nu}\})$ The probability of the first observation time after crashing
Player 1	Decision maker to do the preliminary action
Player 2	The system that might be observed at τ_{ν} after crashing

Based on the above details (Table 7.2), the payoff matrix could be composed and the strategic decision is chosen accordingly from (7.3).

7.3 Defense Operation for Network System

This section provides practical applications to demonstrate how this research could be applied into cyber attack situations. The case deals with the operational strategy for defense operations from the attack. Combating cyber crime involves various mathematical modeling and methods, as well as those of non-mathematical nature. Thus, a few area of current research on this topic and some literature known to the authors. Game and the fluctuation theories are an important hybrid framework and modeling of interest [2, 4]. A router is a network system which is commonly used for connecting servers and the Internet. The hitless-restart mode, as a function of a defense system, which is a special operation mode of a router to be protected from a DoS attack could stay on the forwarding path and the network topology remains stable. In the hitless-restart mode, a router could even restart the entire system software without any shutdown [3]. But this mode could be run only in the limited situation because the router in the hitless-restart mode could not be fully functional (i.e., forwarding only). More importantly, this defense mode should be executed before the router crashes. Thus, forecasting the moment of the system damage is vital to use the hitless-restart mode properly. The hitless-prediction router which capable to predict the moment of queue overflow and takes the preliminary operation, such as nonstop forwarding before the flow hits queue size limitation has been proposed [5]. The hitless-prediction router could choose the mode either by keeping the regular mode or switching the hitless restart mode automatically. The decision making parameters could be the average recycle time, the overflow (or crashing) probabilities, and the payoff values. As it is assumed, the observation process is exponentially distributed with the average observation time $\tilde{\chi}$. The queue size of input stream (C_k) at τ_k is a Poisson process with the parameter λ_0 and the service time process is exponentially distributed with the mean $(1/\rho)$, As such (7.5) can be written as follows:

$$\mu(z, \theta) = \frac{1}{1 + \tilde{\chi}(\lambda - \lambda z + \theta)}, \tag{7.21}$$

where

$$\lambda := \lambda_0 - \rho.$$

As previously mentioned, the initial observation period (7.6) could have different parameters because the initial stream could have existed at the beginning:

$$\mu_0(z, \theta) = \frac{1}{1 + \tilde{\chi}(\lambda_0 - \lambda_0 z + \theta)}. \tag{7.22}$$

If we know the observation index ν after the system crashes, the optimal hitless prediction point is the $(\nu - 1)$th observation point that is one step before the first exceed point. In this case, we find the average index that indicates queue overflows. From the previous research [5] and (7.19),

$$\bar{\nu} = \frac{(S+1)}{\tilde{\chi}\lambda} + \frac{1 - (\lambda_0/\lambda)}{\left(1 + \tilde{\chi}\lambda_0\right)} \cdot \frac{1 - \left[\tilde{\chi}\lambda_0/\left(1 + \tilde{\chi}\lambda_0\right)\right]^{S+1}}{1 - \left[\tilde{\chi}\lambda_0/\left(1 + \tilde{\chi}\lambda_0\right)\right]}, \tag{7.23}$$

and

$$\Gamma^*_{\text{Idle}} = \mathbb{P}\{\nu \geq \bar{\nu}\} = \sum_{k \geq \bar{\nu}} \gamma_k, \tag{7.24}$$

where

$$\gamma_k = \left[\frac{\tilde{\chi}\lambda}{1 + \tilde{\chi}\lambda_0}\right] \left(\frac{1}{1 + \tilde{\chi}\lambda}\right)^k \cdot \left[\sum_{m=0}^{S} \binom{k-1+m}{m} \left(\frac{\left[\tilde{\chi}\lambda_0\right]\left[\tilde{\chi}\lambda\right]}{\left(1 + \tilde{\chi}\lambda_0\right)\left(1 + \tilde{\chi}\lambda\right)}\right)^m\right]. \tag{7.25}$$

It is noted that Γ^*_{Idle} could be the indicator for the optimal pay off ratio Γ^*_{Payoff} from (7.2). Alternatively, the average recycle period of the hitless restart $\bar{\tau}_{\nu-1}$ ($:= \mathbb{E}[\tau_{\nu-1}]$), which is the moment of decision making whether hitless-restart mode or not, is:

$$\bar{\tau}_{\nu-1} = \tilde{\mu}_0 + \tilde{\mu}\mathbb{E}[\nu - 1], \tag{7.26}$$

where

$$\tilde{\mu}_0 = -\frac{\partial \mu_0(1, \theta)}{\partial \theta}\bigg|_{\theta=0}, \tilde{\mu} = -\frac{\partial \mu(1, \theta)}{\partial \theta}\bigg|_{z=0}.$$

Table 7.3 The payoff matrix for hitless-restart routing operations

Unit: [USD]	Regular flow	Overflow (shutdown)
Regular routing mode	0	−30,000
Hitless-restart mode	−8000	−8000

From [5] and (7.26), the recycling period between the starting point and the hitless-prediction point is

$$
\bar{\tau}_{\nu-1} = \tilde{\mu}_0 + \tilde{\mu} \cdot \left[\frac{(S+1)}{\tilde{\chi}\lambda} - 1 + \frac{1 - (\lambda_0/\lambda)}{\left(1 + \tilde{\chi}\lambda_0\right)} \cdot \frac{1 - \left[\tilde{\chi}\lambda_0/\left(1 + \tilde{\chi}\lambda_0\right)\right]^{S+1}}{1 - \left[\tilde{\chi}\lambda_0/\left(1 + \tilde{\chi}\lambda_0\right)\right]} \right].
$$
(7.27)

According to the research [6], the operational cost of a router for typical critical service is USD 30,000 for each shutdown time because of a cyber attack. Before shutdown, the router could be protected by the hitless-restart mode for the defense operation. Because of potential business losses, the defense operation could be increased to USD 8000 in the case of the cost of the business operations under the defense mode.

According to the payoff matrix for the router operation modes (Table 7.3), the best strategy s_1^* for the router operations at the moment $\bar{\tau}_{\nu-1}$ is as follows:

$$
s_1^* = \begin{cases} \text{"Regular routing mode"}, & \Gamma^* \leq \frac{8}{30}, \\ \text{"Hitless-restart mode"}, & \Gamma^* > \frac{8}{30}. \end{cases}
$$
(7.28)

Consequently, the best strategy at the moment of $\bar{\tau}_{\nu-1}$ is determined by the payoff matrix (Table 7.3) and the probability of the router overflow from (7.25).

7.4 Simulated Results

In the simulation, two core functional router systems with the 1 GB capacity that deal with 80% of the forwarding flow and 20% of the routing flow are considered for the comparison. One system has the algorithm for the hitless prediction (i.e., the defense operations) and the other system is a regular router which does not have the defense operation algorithm. The trials are continued until the data size of the queue is exceeding the limit. The regular router could not take any action to avoid the crash of the system. The hitless-prediction router might take the action which is called the defense mode. Unlike the shutdown cost of typical routers which is around 30,000 USD, we assume that the cost of crash is 75,000 USD per an hour and the repair (and reboot) takes 1 h in any status [6] because the cost of the shutdown for a core functional router is very high. The system with the hitless-prediction mode could be over provisioned but we assume that there is no additional cost for over

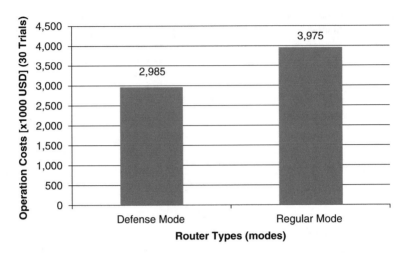

Fig. 7.1 Simulation comparison graph

provisioning (i.e., the cost for the defense operation only). The cost for running defense operations is assumed 15,000 USD per an hour because the router can still forward the flows during the repair time (or the rebooting time). The efficiency of the hitless-prediction router ε that compares to the cost of a regular router can be solved as follows:

$$\varepsilon = \frac{C_d - C_r}{C_r}, \tag{7.29}$$

where C_d is the cost of the defense operation mode and C_r is the crash cost of a regular mode.

After 30 trials, the average cost of the hitless-prediction router which has the defense operation mode is 995,000 USD (2,985,000 USD per 30 trials) and a regular router (without the defense operation mode) costs 132,500 USD (3,975,000 USD per 30 trials). From (7.29), the comparison graph (Fig. 7.1) also shows that the efficiency of the hitless-prediction router is 0.249 that means the hitless-prediction router saves about 25% more than the cost of a regular router.

7.5 Conclusion

The objective of this paper is establishing the theoretical framework and the explicit equations of the mixed strategic decision model with the hybrid of the two-person game and the basic fluctuation theory. This research makes possible estimating the best time for decision making and the mixed strategies to take preliminary defense operations. So, the player could choose the best strategy accordingly. Additionally, the special case supports the general explicit equation for any observation processes with memoryless properties. This analytic approach supports the theoretical background of the decision making time and strategic choice to prevent the crash of the system. This simple framework could be also applied in any practical case of disaster preparedness, stock markets besides a cyber attack defense.

References

1. Dshalalow, J.H.: First excess level process. Advances in Queueing, pp. 244–261. CRC Press, Boca Raton (1995)
2. Dshalalow, J.H., Liew, A.: On fluctuations of a multivariate random walk with some applications to stock options trading and hedging. Math. Comput. Model. **44**(10), 931–944 (2006)
3. Graceful OSPF Restart, IETF Draft. https://tools.ietf.org/html/rfc3623
4. Hida, T. (ed.): Mathematical approach to fluctuations: astronomy, biology and quantum dynamics. In: The Iias Workshop, Kyoto, 18–21 May 1992. World Scientific Publishers, Hackensack (1995)
5. Kim, S.-K.: Design of stochastic hitless-prediction router by using the first exceed level theory. Math. Methods Appl. Sci. **28**(12), 1481–1490 (2005)
6. Oltsik, J.: Best Practices For Availability and Performance Management, Hype-Free Consulting Annual Report (2002)
7. Understanding Denial-of-Service Attacks, Security Tip (ST04-015). https://www.us-cert.gov/ncas/tips/ST04-015

Chapter 8
Optimal Impulse Control of SIR Epidemics Over Scale-Free Networks

Vladislav Taynitskiy, Elena Gubar, and Quanyan Zhu

8.1 Introduction

Malware spreading becomes a more prevalent issue recently as the number of devices and their connections grow exponentially. Many devices that are connected to the Internet do not have strong protections, and they contain cyber vulnerabilities that create a fast spreading of malware over large networks. A higher level of connectivity of the network is often desired for information spreading. However, in the context of malware, the high connectivity can exacerbate the spreading and makes the containment and control of the malware more challenging. One example is the recent Ransomware [10, 11] that spreads over the Internet with the objective to lock the files of a victim using cryptographic techniques and demand a ransom payment to decrypt them. The worldwide spread of WannaCry ransomware has affected more than 200,000 computers across 150 countries and caused billions of dollars of damages. Hence it is critical to take into account the network structure when developing control policies to control the infection dynamics.

In this paper, we investigate a continuous-time Susceptible-Infected-Recovered (SIR) epidemic model over large-scale networks. The malware control mechanism is to patch an optimal fraction of the infected nodes at discrete points in time. Such mechanism is also known as an optimal impulse controller. The hybrid nature of

V. Taynitskiy · E. Gubar
St. Petersburg State University, Faculty of Applied Mathematics and Control Processes,
Saint-Petersburg, Russia
e-mail: e.gubar@spbu.ru

Q. Zhu (✉)
Department of Electrical and Computer Engineering, Tandon School of Engineering,
New York University, Brooklyn, NY, USA
e-mail: quanyan.zhu@nyu.edu

© Springer Nature Switzerland AG 2019
J. B. Song et al. (eds.), *Game Theory for Networking Applications*,
EAI/Springer Innovations in Communication and Computing,
https://doi.org/10.1007/978-3-319-93058-9_8

discrete-time control policy of continuous-time epidemic dynamics together with the network structure poses a challenging optimal control problem. We leverage the Pontryagin's minimum principle for impulsive systems to obtain an optimal structure of the controller and use numerical experiments to demonstrate the computation of the optimal control and the controlled dynamics. This work extends the investigation of previous related works [7, 9, 15] to a new paradigm of coupled epidemic models and the regime of optimal impulsive control.

The rest of the paper is organized as follows. Section 8.2 presents the controlled SIR mathematical model. Section 8.3 shows the structure of optimal control policies. Section 8.4 presents numerical examples. Section 8.5 concludes the paper.

8.2 The Model

In this section, we formulate a model to describe the spreading of malware in the network of N nodes by extending the classical SIR model. As in previous works [7, 14] two different forms of malware with different strengths spread over the network simultaneously, we denote them as M_1 and M_2. We also assume that a structure of population is described by the scale free network [8, 12]. Normally, as SIR model points, all nodes in the population are divided into three groups: *Susceptible* (*S*), *Infected* (*I*), and *Recovered* (*R*). *Susceptible* is a group of nodes which are not infected by any malware, but may be invaded by any forms of virus. The *Infected* nodes are those that have been attacked by the virus and the *Recovered* is a group of recovered nodes. In modified model subgroup of Infected nodes also is brunched into two subgroups I_1 and I_2, where nodes in I_i are infected by malware $M_i, i = 1, 2$, respectively. We formulate the epidemic process as a system of nonlinear differential equations, where n_S, n_{M_1}, n_{M_2}, and n_R correspond to the number of susceptible, infected, and recovered nodes, respectively. In current model the connections between nodes are described by the scale-free network, then we will use the following notation: $S_k(t)$ and $R_k(t)$ are fractions of *Susceptible* and *Recovered* nodes with degree k at time moment t, $I_k^1(t)$, $I_k^2(t)$ are fractions of *Infected* nodes with degree k. At each time moment $t \in [0, T]$ the number of nodes is constant and equal N, and the following condition $S_k(t) + I_k^1(t) + I_k^2(t) + R_k(t) = 1$ is satisfied. The process of spreading is defined by the system of ordinary differential equations:

$$
\begin{aligned}
\frac{dS_k}{dt} &= -\delta_{1k} S_k I_k^1 \Theta_1 - \delta_{2k} S_k I_k^2 \Theta_2; \\[2mm]
\frac{dI_k^1}{dt} &= \delta_{1k} S_k I_k^1 \Theta_1 - \sigma_k^1 I_k^1; \\[2mm]
\frac{dI_k^2}{dt} &= \delta_{2k} S_k I_k^2 \Theta_2 - \sigma_k^2 I_k^2, \\[2mm]
\frac{dR_k}{dt} &= \sigma_k^1 I_k^1 + \sigma_k^2 I_k^2,
\end{aligned}
\tag{8.1}
$$

where $\delta_{ik}(k)$ is the infections rate for the first type of malware i if a susceptible node has a contact with infected node with the degree k, σ_k^i is recovery rate.

We consider the graph generated by using the algorithm devised in [2]. We start from a small number m_0 of disconnected nodes; every time step a new node is added, with m links that are connected to an old node i with k_i links according to the probability $k_i / \sum_j k_j$. After iterating this scheme a sufficient number of times, we obtain a network composed by N nodes with connectivity distribution $P(k) \approx k^{-3}$ and average connectivity $\langle k \rangle = 2m$. In this work we take $m = 4$.

At the initial time moment $t = 0$, the most number of nodes belong to Susceptible group and only a small fraction of Infected by malwares M_1 or M_2. Initial state for system (8.1) is $0 < S_k(0) < 1$, $0 < I_k^1(0) < 1$, $0 < I_k^2(0) < 1$, $R(0) = 1 - S_k(0) - I_k^1(0) - I_k^2(0)$. Analogously with [6, 12] we define parameter $\Theta_i(t)$ as

$$\Theta_i(\lambda_i) = \sum_{k'} \frac{\delta_{ik} P(k'|k) I_{k'}^i}{k'}, \quad i = 1, 2, \tag{8.2}$$

where δ_{ik} denotes the infectivity of a node with degree k and $\lambda_i = \delta_{ik}/\sigma_k^i$ an effective spreading rate. $P(k'|k)$ describes the probability of a node with degree k pointing to a node with degree k', and $P(k'|k) = \frac{k'P(k')}{\langle k \rangle}$, where $\langle k \rangle = \sum_{k'} k P(k)$. For scale-free node distribution $P(k) = C^{-1} k^{-2-\gamma}$, $0 < \gamma \le 1$, where $C = \zeta(2+\gamma)$ is Riemann's zeta function, which provides an appropriate normalization constant for sufficiently large networks. In the SF model considered here, we have a connectivity distribution $P(k) = 2m^2/k^{-3}$, where k is approximated as a continuous variable. According to [12] we can rewrite (8.2) as

$$\Theta_i(\lambda_i) = \frac{e^{-1/m\lambda_i}}{m\lambda_i}, \quad i = 1, 2. \tag{8.3}$$

8.3 Impulse Control Problem

Previously it was shown in [12] a small fraction of the infected nodes might be survived on small segments of the network and can provoke new waves of epidemics. This cycled process recalls the behavior of the virus of influenza which causes a seasonally periodic epidemic [1]. Hence the control of the epidemic process can be formulated as an impulse control problem in which a series of impulses of antivirus patches are designed to reduce the periodically incipient zones of infected nodes. We extend the model (8.1) to present an impulse control problem for episodic attacks of the malware and obtain the optimal strategy of application of antivirus software to damp the spreading of malware at discrete time moments.

We suppose that impulses occur at time $\tau_{k,1}^i, \ldots, \tau_{k,q_i(k)}^i$, where $q_i(k)$ describes the number of launching of impulse controls for nodes with k degrees, index i indicates the type of malware. We also assume that on the time intervals $(\tau_{k,j}^i, \tau_{k,j+1}^i]$ system (8.1) describes the behavior of malware in the network. We have reformulated epidemic model to describe the situation with two types of malware for all time periods except the sequence of times $\tau_{k,j}^{i+}$, $j = 1, \ldots, q_i(k)$, $i = 1, 2$. Additionally, we set $S(\tau_{k,j}^i) = S(\tau_{k,j}^{i-})$, $I_1(\tau_{k,j}^i) = I_1(\tau_{k,j}^{i-})$, $I_2(\tau_{k,j}^i) = I_2(\tau_{k,j}^{i-})$, $R(\tau_{k,j}^i) = R(\tau_{k,j}^{i-})$.

The system after activation of impulses at time moment $\tau_{k,j}^{i+}$ is:

$$
\begin{aligned}
S_k\left(\tau_{k,j}^{i+}\right) &= S_k\left(\tau_{k,j}^i\right), \\
I_k^1\left(\tau_{k,j}^{i+}\right) &= I_k^1\left(\tau_{k,j}^i\right) - v_k^1\left(\tau_{k,j}^i\right), \\
I_k^2\left(\tau_{k,j}^{i+}\right) &= I_k^2\left(\tau_{k,j}^i\right) - v_k^2\left(\tau_{k,j}^i\right), \\
R_k\left(\tau_{k,j}^{i+}\right) &= R_k\left(\tau_{k,j}^i\right) + v_k^1\left(\tau_{k,j}^i\right) + v_k^2\left(\tau_{k,j}^i\right).
\end{aligned}
\tag{8.4}
$$

Variables $v_k^i = (v_{k,1}^i, \ldots, v_{k,q_i(k)}^i)$, $i = 1, 2$, correspond to control impulses applied at the discrete time moments $\tau_{k,1}, \ldots, \tau_{k,q_i(k)}$ and represent the fraction of recovered nodes. Let be $v_{k,j}^i = c_{k,j}^i \delta(t - \tau_{k,j}^i)$, where $\delta(t - \tau_{k,j}^i)$ is Dirac function, $c_{k,j}^i \in [0, \bar{u}_{k,j}^i]$ is the value of impulse, leads to changes of the dynamical system, $\bar{u}_{k,j}^i$ is the maximum value for control [1].

Functional The objective function of the combined system (8.4) is represented by the aggregated costs on the time interval $[0, T]$ including the costs of control impulses. The aggregated costs for continuous system (8.1) are defined as follows: at time moment $t \neq \tau_{k,j}^i$, $j = 1, \ldots, q_i(k)$, $i = 1, 2$, we have the costs from infected nodes $f_k^1(I_k^1(t))$ and $f_k^2(I_k^2(t))$. Functions $f_k^i(\cdot)$ are non-decreasing and twice-differentiable, such that $f_k^i(0) = 0$, $f_k^i(I_k^i(t)) > 0$ for $I_k^i(t) > 0$ with $t \in (\tau_{k,j-1}^i, \tau_{k,j}^i]$. For system (8.4), we define the treatment costs as functions $h_k^i(v_{k,j}^i(\tau_{k,j}^{i+}))$, $j = 1, \ldots, q_i(k)$, where $h_k^i(v_{k,j}^i(\tau_{k,j}^{i+})) > 0$, $v_{k,j}^i(\tau_{k,j}^{i+}) > 0$ for $i = 1, 2$. Functions $g(R_k(t))$ are non-decreasing and capture the benefit rates from Recovered nodes. The aggregated system costs are defined by the functional:

$$
\begin{aligned}
J = \sum_{k \in \mathbb{N}} \Bigg[&\int_0^T f_k^1\left(I_k^1(t)\right) + f_k^2\left(I_k^2(t)\right) - g\left(R(t)\right) dt \\
&+ \sum_{j=1}^{q_1(k)} h_k^1\left(v_{k,j}^1(\tau_{k,j}^1)\right) + \sum_{j=1}^{q_2(k)} h_k^2\left(v_{k,j}^2(\tau_{k,j}^2)\right) \Bigg].
\end{aligned}
\tag{8.5}
$$

8.4 The Structure of Impulse Control

According to principle maximum in impulse form [3–5, 15] we write Hamiltonian for dynamic system (8.1)

$$
\begin{aligned}
H_k^0(t) = &-\left(f_k^1\big(I_k^1(t)\big) + f_k^2\big(I_k^2(t)\big) - g\big(R_k(t)\big) \right) \\
&+ \big(\lambda_{I_k^1}(t) - \lambda_{S_k}(t)\big)\delta_{1k} S_k(t) I_k^1(t)\Theta_1(t) \\
&+ \big(\lambda_{I_k^2}(t) - \lambda_{S_k}(t)\big)\delta_{2k} S_k(t) I_k^2(t)\Theta_2(t) + \big(\lambda_{R_k} - \lambda_{I_k^1}\big)\sigma_k^1 I_k^1 \\
&+ \big(\lambda_{R_k} - \lambda_{I_k^2}\big)\sigma_k^2 I_k^2;
\end{aligned}
\tag{8.6}
$$

and construct adjoint system as follows:

$$
\begin{aligned}
\dot{\lambda}_{S_k}(t) &= \big(\lambda_{S_k}(t) - \lambda_{I_k^1}(t)\big)\delta_{1k} I_k^1(t)\Theta_1(t) + \big(\lambda_{S_k}(t) - \lambda_{I_k^2}(t)\big)\delta_{2k} I_k^2(t)\Theta_2(t); \\
\dot{\lambda}_{I_k^1}(t) &= \frac{d f_k^1\big(I_k^1(t)\big)}{d I_k^1} \\
&\quad + \big(\lambda_{S_k}(t) - \lambda_{I_k^1}(t)\big)\left(\delta_{1k} S_k(t)\Theta_1(t) + \frac{(\delta_{1k})^2 S_k(t) I_k^1(t) P(k)}{\langle k \rangle}\right) \\
&\quad + \big(\lambda_{I_k^1} - \lambda_{R_k}\big)\sigma_k^1; \\
\dot{\lambda}_{I_k^2}(t) &= \frac{d f_k^2\big(I_k^2(t)\big)}{d I_k^2} \\
&\quad + \big(\lambda_{S_k}(t) - \lambda_{I_k^2}(t)\big)\left(\delta_{2k} S_k(t)\Theta_2(t) + \frac{(\delta_{2k})^2 S_k(t) I_k^2(t) P(k)}{\langle k \rangle}\right) \\
&\quad + \big(\lambda_{I_k^2} - \lambda_{R_k}\big)\sigma_k^2; \\
\dot{\lambda}_{R_k}(t) &= -\frac{d g\big(R_k(t)\big)}{d R_k},
\end{aligned}
\tag{8.7}
$$

with transversality conditions $\lambda_{S_k}(T) = \lambda_{I_k^1}(T) = \lambda_{I_k^2}(T) = \lambda_{R_k}(T) = 0$.

Following the maximum principle for impulse control (see [3, 4, 13]), we formulate necessary optimality conditions as in Theorem 1.

Theorem 1 *Let* $(x^*, N, \tau_i^{j*}, v_i^*)$, $i = 1, 2$, *be an optimal solution for the impulse control problem. Then, there exists an adjoint vector function* $\lambda(t) = (\lambda_S(t), \lambda_{I_1}(t), \lambda_{I_2}(t), \lambda_R(t))$ *such that the following conditions hold:*

$$
\dot{\lambda}_x(t) = -\frac{\partial H_0}{\partial x}(x^*(t), \lambda(t), t),
\tag{8.8}
$$

where $x(t) = S(t), I_1(t), I_2(t), R(t)$.

At the impulse or jump points, it holds that

$$\frac{\partial H_i^c}{\partial v_i}\left(x^*\left(\tau_i^{j*-}\right), v_i, \lambda\left(\tau_i^{j*+}\right), \tau_i^{j*}\right)\left(v_i^j - v_i^{j*}\right) \geq 0,$$ (8.9)

$$\lambda_x\left(\tau_i^{j*+}\right) - \lambda_x\left(\tau_i^{j*-}\right) = \frac{\partial H_i^c}{\partial x}\left(x^*\left(\tau_i^{j*-}\right), v_i^{j*}, \lambda\left(\tau_i^{j*+}\right), \tau_i^{j*}\right),$$ (8.10)

$$H_0\left(x^*\left(\tau_i^{j*+}\right), \lambda\left(\tau_i^{j*+}\right), \tau_i^{j*}\right) - H_0\left(x^*\left(\tau_i^{j*-}\right), \lambda\left(\tau_i^{j*-}\right), \tau_i^{j*}\right)$$

$$-\frac{\partial H_i^c}{\partial \tau_i^j}\left(x^*\left(\tau_i^{j*-}\right), v_i^{j*}, \lambda\left(\tau_i^{j*+}\right), \tau_i^{j*}\right)\begin{cases} > 0 & for\ \tau_i^{j*} = 0, \\ = 0 & for\ \tau_i^{j*} \in (0, T), \\ < 0 & for\ \tau_i^{j*} = T. \end{cases}$$ (8.11)

For all points in time at which there is no jump, i.e. $t \neq \tau_j$ ($j = 1, \ldots, k_i$), it holds that

$$\frac{\partial H_j^c}{\partial v_j}\left(x^*(t), 0, \lambda(t), t\right)v_j \leq 0,$$ (8.12)

with the transversality condition $\lambda(T) = 0$.

Hamiltonian in impulsive form is

$$H_k^c\left(\tau_{k,j}^{1+}\right) = -h_k^1\left(v_{k,j}^1\left(\tau_{k,j}^{1+}\right)\right) + \left(\lambda_{R_k}(\tau_{k,j}^{1+}) - \lambda_{I_k^1}(\tau_{k,j}^{1+})\right)v_{k,j}^1\left(\tau_{k,j}^{1+}\right);$$

$$H_k^c\left(\tau_{k,j}^{2+}\right) = -h_k^2\left(v_{k,j}^2\left(\tau_{k,j}^{2+}\right)\right) + \left(\lambda_{R_k}(\tau_{k,j}^{2+}) - \lambda_{I_k^2}(\tau_{k,j}^{2+})\right)v_{k,j}^2\left(\tau_{k,j}^{2+}\right).$$ (8.13)

Here we assume that for each type of malwares M_1 and M_2 and for each k we have own set of control impulses $v_k^1 = (v_{k,1}^1, \ldots, v_{k,q_1(k)}^1)$ and $v_k^2 = (v_{k,1}^2, \ldots, v_{k,q_2(k)}^2)$.

Adjoin system for system(8.4) is ($i = 1, 2$):

$$\lambda_{S_k}\left(\tau_{k,j}^{i+}\right) = \lambda_{S_k}\left(\tau_{k,j}^i\right);$$

$$\lambda_{I_k^1}\left(\tau_{k,j}^{i+}\right) = \lambda_{I_k^1}\left(\tau_{k,j}^i\right);$$ (8.14)

$$\lambda_{I_k^2}\left(\tau_{k,j}^{i+}\right) = \lambda_{I_k^2}\left(\tau_{k,j}^i\right);$$

$$\lambda_{R_k}\left(\tau_{k,j}^{i+}\right) = \lambda_{R_k}\left(\tau_{k,j}^i\right).$$

Here are the conditions for Δ_i for each I_k^i from the Theorem 1:

$$\Delta_1 = f_k^1\big(I_k^1(\tau_{k,j}^1)\big) - f_k^1\big(I_k^1(\tau_{k,j}^{1+})\big) - g\big(R_k(\tau_{k,j}^1)\big) + g\big(R_k(\tau_{k,j}^{1+})\big)$$

$$+c_{k,j}^1\left[\frac{dg\big(R_k(\tau_{k,j}^{1+})\big)}{dR_k(\tau_{k,j}^{1+})} + \frac{df_k^1\big(I_k^1(\tau_{k,j}^{1+})\big)}{dI_k^1(\tau_{k,j}^{1+})}\right] + 2\sigma_k^1 c_{k,j}^1\big(\lambda_{R_k}(\tau_{k,j}^{1+}) - \lambda_{I_k^1}(\tau_{k,j}^{1+})\big)$$

$$\delta_{1k} S_k(\tau_{k,j}^1)c_{k,j}^1\big(\lambda_{S_k}(\tau_{k,j}^1) - \lambda_{I_k^1}(\tau_{k,j}^1)\big)\left[2\Theta_1(\tau_{k,j}^{1+}) + \frac{\delta_{1k}P(k)}{\langle k\rangle}\big(1 + I_k^1(\tau_{k,j}^1)\big)\right.$$

$$\left. -c_{k,j}^1\big)\right].$$

(8.15)

Here are the conditions for Δ for each Δ for each I_k^2 from Theorem 1:

$$\Delta_2 = f_k^2\big(I_k^2(\tau_{k,j}^2)\big) - f_k^2\big(I_k^2(\tau_{k,j}^{2+})\big) - g\big(R_k(\tau_{k,j}^2)\big) + g\big(R_k(\tau_{k,j}^{2+})\big)$$

$$+c_{k,j}^2\left[\frac{dg\big(R_k(\tau_{k,j}^{2+})\big)}{dR_k(\tau_{k,j}^{2+})} + \frac{df_k^2\big(I_k^2(\tau_{k,j}^{2+})\big)}{dI_k^2(\tau_{k,j}^{2+})}\right] + 2\sigma_k^2 c_{k,j}^2\big(\lambda_{R_k}(\tau_{k,j}^{2+}) - \lambda_{I_k^2}(\tau_{k,j}^{2+})\big);$$

$$\delta_{2k} S_k(\tau_{k,j}^2)c_{k,j}^2\big(\lambda_{S_k}(\tau_{k,j}^2) - \lambda_{I_k^2}(\tau_{k,j}^2)\big)\left[2\Theta_2(\tau_{k,j}^{2+}) + \frac{\delta_{2k}P(k)}{\langle k\rangle}\big(1 + I_k^2(\tau_{k,j}^2)\big)\right.$$

$$\left. -c_{k,j}^2\big)\right].$$

(8.16)

According to Theorem 1 at time $\tau_{k,j}^i \in (0,T)\Delta_i$ should be equal to zero. Therefore, we deal with two different problems: firstly, if the intensity of impulses $c_{k,j}^i$ are fixed, then from (8.15) and (8.16), we can find the optimal time $\tau_{k,j}^{i*}$ of using impulses; secondly, if the sequence of time $\tau_{k,j}^i$ are fixed, then we obtain the optimal level of the intensity of impulses $c_{k,j}^{i*}$, $j = 1, \ldots, q_i$, $i = 1, 2$.

8.5 Numerical Simulations

In this paragraph we present numerical experiments to depict theoretical results and to study the behavior of malwares and show the application of control impulses. Here we use the following set of the initial states and values of parameters of the system (8.1): initial system states and parameters are $S_k(0) = 0.4$, $I_k^1(0) = 0.3$, $I_k^2(0) = 0.2$, and $R_k(0) = 0$, spreading rates are $\delta_{1k} = 0.075k$, $\delta_{2k} = 0.1k$, self-recovery rates are $\sigma_k^1 = 0.0005k$ and $\sigma_k^1 = 0.0003k$. We set costs functions for infectious subgroups as $f_k^i(I_k^i(t)) = A_k^i I_k^i(t)$ with coefficients $A_k^1 = 2k$, $A_k^2 = 3k$ and treatment costs functions as $h_k^i(v_{k,j}^i(\tau_{k,j}^{i+})) = B_k^i c_{k,j}^i I_k^i(\tau_{k,j}^{i+})$, where coefficients are equal to $B_k^1 = 3k$, $B_k^2 = 4k$, $c_{k,j}^1 = 0.1$, $c_{k,j}^2 = 0.08$ for $i = 1, 2$, utility function is $g(R_k(t)) = 0.1R_k(t)$.

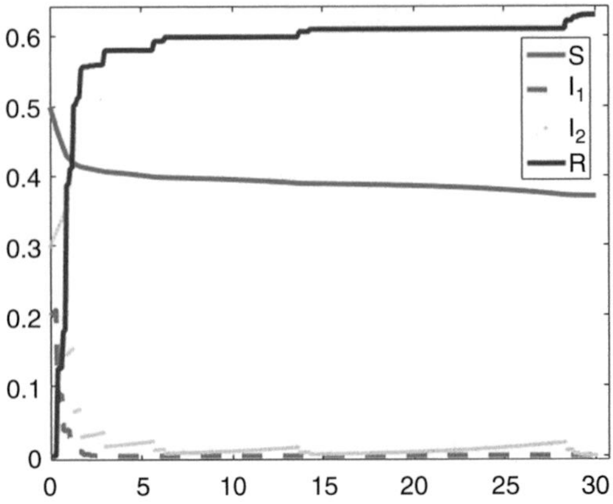

Fig. 8.1 Evolution of the
system in Case 1. Number of
links: $k = 4$, spreading rates:
$\delta_{1k} = 0.075k$ and $\delta_{2k} = 0.1k$

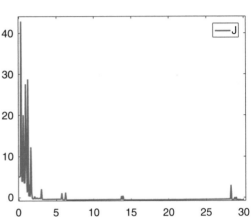

Fig. 8.2 Aggregated system
costs are equal to $J = 37.65$

Case 1 In case 1 we present the initial example of the behavior of the system and
aggregated system costs if an average number of links between ith node and its
neighbors is $k = 4$. Figures 8.1 and 8.2 show the spreading on two modification of
malwares and corresponding total system costs.

Aggregated system costs in this case are equal to $J = 37.65$. By applying the
control impulses at discrete time moments we received that an amount of impulses
are equal to $p_1(4) = 37$ and $p_2(4) = 49$.

Case 2 In this experiment we use the same parameters for initial data, but in
contrast to case 1 an average number on neighbors is equal $k = 7$. In this case,
we obtain that the aggregated costs are $J = 73.93$, and an amount of impulses are
equal to $p_1(7) = 29$ and $p_2(7) = 44$. We may notice that increasing the number
of neighbor links increases the costs of the system. Since there are less nodes with
connectivity $k = 7$ which is more than average connectivity $\langle k \rangle = 4$, we need less
impulse treatment to vaccinate the network, thereby if we apply control to more
connected nodes we reduce the costs of treatment (Fig. 8.3).

Fig. 8.3 (a) Evolution of the system in Case 2. Number of links: $k = 7$, spreading rates: $\delta_{1k} = 0.075k$ and $\delta_{2k} = 0.1k$. (b) Aggregated system costs are equal to $J = 73.93$

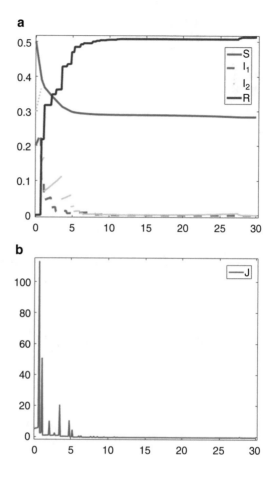

Case 3 In case 3, by using the same initial set of data we variate the spreading rate for malwares and consider $\delta_{1k} = 0.075k$ and $\delta_{2k} = 0.1k$. Here we receive that the aggregated costs are $J = 122.27$ and a number of impulses are $p_1(4) = 43$ and $p_2(4) = 55$, then increasing the spreading rates is leading to increasing aggregated costs and number of impulses which are needed to heal the network (Fig. 8.4).

8.6   Conclusion

This work addresses the spreading of cyber threats over large-scale networks by investigating the optimal control policies in the impulsive form for SIR-type of epidemics over scale-free networks. We have applied the impulse optimal control framework to the epidemics over networks to devise impulsive protection policies to mitigate the impact of the epidemics and contain the spreading of the malware.

Fig. 8.4 (**a**) Evolution of the
system in Case 3. Number of
links: $k = 4$, spreading rates:
$\delta_{1k} = 0.1k$ and $\delta_{2k} = 0.2k$.
(**b**) Aggregated system costs
are equal to $J = 122.27$

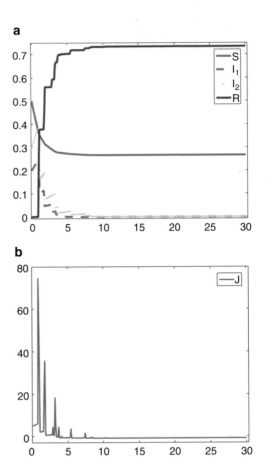

With the application of the maximum principle, we have obtained the structure of
the optimal control impulses where actions are taken at discrete-time moments. We
have corroborated the obtained results using numerical examples.

Acknowledgements The work of the second author was supported by the research grant "Optimal
Behavior in Conflict-Controlled Systems" (17-11-01079) from Russian Science Foundation.

References

1. Agur, Z., Cojocaru, L., Mazor, G., Anderson, R.M., Danon, Y.L.: Pulse mass measles
   vaccination across age cohorts. Proc. Natl. Acad. Sci. USA **90**, 11698–11702 (1993)
2. Barabási, A.L., Albert, R.: Emergence of scaling in random networks. Science **286**(5439),
   509–512 (1999)
3. Blaquière, A.: Impulsive optimal control with finite or infinite time horizon. J. Optim. Theory
   Appl. **46**, 431–439 (1985)

4. Chahim, M., Harti, R., Kort, P.: A tutorial on the deterministic impulse control maximum principle: necessary and sufficient optimality conditions. Eur. J. Oper. Res. **219**, 18–26 (2012)
5. Dykhta, V.A., Samsonyuk, O.N.: A maximum principle for smooth optimal impulsive control problems with multipoint state constraints. Comput. Math. Math. Phys. **49**, 942–957 (2009)
6. Fu, X., Small, M., Walker, D.M., Zhang, H.: Epidemic dynamics on scale-free networks with piecewise linear infectivity and immunization. Phys. Rev. E. **77**(3), 036113 (2008)
7. Gubar, E., Zhu, Q.: Optimal control of influenza epidemic model with virus mutations. In: Proceedings 12th Biannual European Control Conference, pp. 3125–3130. IEEE Control Systems Society, New York (2013)
8. Gubar, E., Kumacheva, S., Zhitkova, E., Porokhnyavaya, O.: Impact of propagation information in the model of tax audit. In: Recent Advances in Game Theory and Applications. Static and Dynamic Game Theory: Foundations and Applications, Switzerland, pp. 91–110 (2015)
9. Gubar, E., Zhu, Q., Taynitskiy, V.: Optimal control of multi-strain epidemic processes in complex networks. In: Game Theory for Networks. GameNets 2017. Lecture Notes of the Institute for Computer Sciences, Social Informatics and Telecommunications Engineering, vol. 212, pp. 108–117. Springer, Cham (2017)
10. Kharraz, A., Robertson, W., Balzarotti, D., Bilge, L., Kirda, E.: Cutting the gordian knot: a look under the hood of ransomware attacks. In: International Conference on Detection of Intrusions and Malware, and Vulnerability Assessment, pp. 3–24. Springer, Berlin (2015)
11. Luo, X., Liao, Q.: Ransomware: a new cyber hijacking threat to enterprises. In: Handbook of Research on Information Security and Assurance, pp. 1–6. IGI Global, Hershey (2009)
12. Pastor-Satorras, R., Vespignani A.: Epidemic spreading in scale-free networks. Phys. Rev. Lett. **86**(14), 3200 (2001)
13. Sethi, S.P., Thompson, G.L.: Optimal Control Theory: Applications to Management Science and Economics. Springer, Berlin (2006)
14. Taynitskiy, V.A., Gubar, E.A., Zhitkova, E.M.: Optimization of protection of computer networks against malicious software. In: Proceedings of International Conference Stability and Oscillations of Nonlinear Control Systems (Pyatnitskiy's Conference) (2016)
15. Taynitskiy, V., Gubar, E., Zhu Q.: Optimal impulse control of bi-virus SIR epidemics with application to heterogeneous internet of things. In: Constructive Nonsmooth Analysis and Related Topics. Abstracts of the International Conference. Dedicated to the Memory of Professor V.F. Demyanov, pp. 113–116 (2017)

Part II
Game Theory and Social Network Analysis

Chapter 9
Fault-Tolerant Hotelling Games

Chen Avin, Avi Cohen, Zvi Lotker, and David Peleg

9.1 Introduction

Background The Hotelling game, originated in Hotelling's seminal work in 1929 [9], modeled the competition of two servers on a linear market (e.g., two ice-cream vendors on a beach strip) as follows. Two servers choose a location on the line segment [0, 1], and the payoff of each server is equal to the length of the segment of points closer to it than the other server (a.k.a. its Voronoi cell). Hotelling showed that if both servers locate themselves at the center of the line, then a Nash equilibrium would be reached, i.e., a situation in which neither server would rather relocate unilaterally.

He next showed that three servers competing on the line reach no equilibrium state—in every configuration of servers there will be one server who can increase its profit by moving.

This initial idea sparked decades of research. Notably, in 1975, Eaton and Lipsey [5] completely characterized all Nash equilibria of the n server game, for every n. Over the years numerous variations were made to each component of the game, including the number of players, the pricing policy, the behavior of clients, and

C. Avin (✉) · Z. Lotker
Ben-Gurion University of the Negev, Beer-Sheba, Israel
e-mail: avin@cse.bgu.ac.il; zvilo@cse.bgu.ac.il

A. Cohen · D. Peleg
Weizmann Institute of Science, Rehovot, Israel
e-mail: avi.cohen@weizmann.ac.il; david.peleg@weizmann.ac.il

© Springer Nature Switzerland AG 2019
J. B. Song et al. (eds.), *Game Theory for Networking Applications*,
EAI/Springer Innovations in Communication and Computing,
https://doi.org/10.1007/978-3-319-93058-9_9

the geometry of the market. Eiselt et al. [7] provide an annotated bibliography categorized by these features (for a more recent survey see Eiselt et al. [6]).

Motivation In this paper, we consider the Hotelling game in a failure prone setting, and explore how the Hotelling game behaves differently in a world where faults may occur either in the environment or in the servers themselves. In day to day life, uncertainly must be accounted for. For instance, one of the players may fail to open their store due to illness or vacation, or the road might be blocked due to infrastructure work or safety issues, denying clients access to their preferred vendor. While it is uncertain whether such an event occurs on a given day or not, it is certain to happen one day. It therefore stands to reason that failures would be accounted for in player strategies and payoffs. Indeed, fault-tolerant problems constitute a fertile area of research in Computer Science. Yet, to the best of our knowledge, this paper is the first to consider fault-tolerance aspects of the Hotelling game.

Contributions We analyzed two types of failure models. In the first variant, we consider a failure prone *environment*, wherein it is possible that the line would be blocked at a random point, denying the passage of clients through it. We refer to this variant of the game with n players as the *Line Failure Hotelling game* and denote it by $H_{\mathrm{LF}}(n)$. We characterize all existing equilibria of $H_{\mathrm{LF}}(n)$ (Theorem 3). Moreover, we show that each Nash equilibrium of $H_{\mathrm{LF}}(n)$ corresponds to a Nash equilibrium of the non-failure Hotelling game $H(n)$ (Theorem 2).

The second variant we consider assumes a reliable environment, but failure prone players. Each player (independently) has some probability of being removed from the game. This version of the game (with n players) is referred to as the *Player Failure Hotelling game*, denoted $H_{\mathrm{PF}}(n)$. We show that for $n \geq 3$ players, the game admits no Nash equilibrium (Theorem 4).

Related Work Fault tolerant facility location problems have been studied extensively from an optimization perspective in the operations research and computer science communities [2, 3, 10–13]. However, relatively little work has been done from a game theoretic approach. Two recent papers do consider fault tolerant location games. Wang and Ouyang studied a failure prone competitive location game in a two-dimensional environment [14]. However, they studied a variant in which two players position several facilities each, and did not extend the model to a greater number of players. Zhang et al. considered a discrete competitive location model in which there are finitely many clients and finitely many potential facility locations [15]. Their model, while similar to our model in theme, bears little resemblance to Hotelling's original game.

In 1987, De Palma et al. [4] introduced randomness into the Hotelling game, though in a manner different than we do. Rather than uncertainty in the reliability of the environment or the players, their paper studied uncertainty in client behaviors. That is, clients normally shop at the closest vendor, but with some probability, due to unquantifiable factors of personal taste, they skip a seller and travel further to the next one. This paper has a similar approach to our paper, but considers a different problem than ours.

Other types of fault tolerant games have recently been studied. For example, Gradwohl and Reingold [8] studied the immunity and tolerance of games with many players. Immunity means that faults have a small affect on the utility of non-faulty players. Tolerance means that optimal strategies remain optimal when faults are introduced to the game (even though the utilities may be different from the case without faults). The authors show that the *games themselves* are resilient to faults and quantify the strength of their resilience. We, on the other hand, consider a game that is very sensitive to faults and ask how would the players adapt their strategies in a given fault model.

Organization The paper is organized as follows. Section 9.2 presents the Hotelling game formally along with known results. Sections 9.3 and 9.4 analyze the Hotelling game with line disconnections and server crashes, respectively. Section 9.5 concludes the paper and offers future directions of research.

9.2   Model and Preliminaries

The Game The Hotelling game on the line segment $[a, b]$, denoted as $H(n, [a, b])$, involves n servers, s_i for $i = 1, \ldots, n$, who place themselves at different points x_i on the segment. Each point on the line also represents a client, who will be served by the nearest server. The *market* of each server is the line segment containing the clients that will be served by it (this segment is also known as the server's *Voronoi cell*). The payoff of each server s_i, denoted $p(s_i)$, is the length of his market. Servers strive to maximize their payoff.

We assume clients are uniformly distributed over the line. We also assume that no more than one server can occupy a given location; the minimal distance between two servers is some arbitrarily small $\varepsilon > 0$. (Setting a minimal distance between servers is common practice in the Hotelling game literature. This prevents servers from infinitely moving closer and closer to each other to slightly improve their payoff. However, the servers choose their location simultaneously and thus two servers might inadvertently choose the same location. We assume that in this case players make small corrections until they meet the constraint. Alternatively, we could say that two servers can be located at the same point and split the payoff in half. However, this leads to unnatural equilibria and thus makes the analysis more complicated. For example, in the two-server game, locating both servers at the same point, anywhere on the line, yields an equilibrium.) When two servers are separated by a distance of ε they are said to be *paired*. We say two servers s_i and s_j are *paired at location* x if $x_i = x - \varepsilon/2$ and $x_j = x + \varepsilon/2$; with a slight abuse of notation, we hereafter denote this as $x_i = x_j = x$. Conversely, a server is *isolated* if it is not paired to another server.

Each server s_i divides its market into two sides, referred to as *half-markets*. We denote by $L(s_i)$ and $R(s_i)$ the lengths of s_i's left and right half-markets, respectively. Therefore, the payoff of s_i is $p(s_i) = L(s_i) + R(s_i)$. Two servers are said to be *neighbors* if no server is located between them. A server that has neighbors on both sides is called an *interior server*. A server that has one neighbor is called a

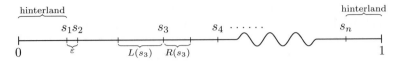

Fig. 9.1 A possible configuration of the game. Servers s_1 and s_n are the peripheral servers, s_1 and s_2 are paired, and the half-markets of s_3 are marked in the figure

peripheral server. That is, the two peripheral servers are the server closest to 0 and the server closest to 1 (See Fig. 9.1).

By definition, the market of an interior server extends half the distance to its two neighbors. The length of an interior server's market is thus half the distance between its neighbors, wherever the server is located between those neighbors. The line segment between the boundary and the corresponding peripheral server is called a *hinterland*. The market of a peripheral server includes its hinterland in one direction, and extends half way to its neighbor in the other direction.

The game is in a *Nash equilibrium* if no server can increase its payoff by moving to a location other than its present location.

Known Results Eaton and Lipsey [5] proved that the following are necessary and sufficient conditions for an equilibrium:

(EL1)   Each peripheral server is paired.
(EL2)   No server's whole market is smaller than any other server's half-market.

The proof that these conditions are sufficient is a bit involved, but it is easy to see why they are necessary. If (EL1) does not hold, a peripheral server would increase its profit by locating closer to its neighbor. (EL2) follows from the fact that any server can obtain a market equal to any other server's half-market by pairing with it.

By applying these equilibrium conditions, Eaton and Lipsey determined the equilibria of the game depending on the number of servers, as follows.

One Server The payoff of a single server s_1 is $p(s_1) = 1$, wherever it is located.

Two Servers As shown by Hotelling, there is a unique equilibrium, where the two servers are paired at $x_1 = x_2 = 1/2$, with payoffs $p(s_1) = p(s_2) = 1/2$.

Three Servers No equilibrium exists. It follows from the fact that (El1) and (El2) contradict one another in this case.

Four Servers There is a unique equilibrium, with equal payoffs of $1/4$, where two servers are paired at $x_1 = x_2 = 1/4$, and the other two at $x_3 = x_4 = 3/4$.

Five Servers There is a unique equilibrium, where two servers are paired at $x_1 = x_2 = 1/6$, two others are paired at $x_4 = x_5 = 5/6$, and an isolated server is located at $x_3 = 1/2$. Note that here, the payoffs are not uniform: s_3 has payoff $p(s_3) = 1/3$, while all other servers have payoff $1/6$.

More than Five Servers There exist infinitely many equilibria, characterized as follows: peripheral servers are paired with their neighbors and have identical

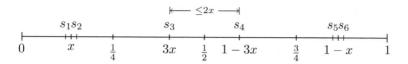

Fig. 9.2 Equilibrium with six servers

hinterlands. Each peripheral pair is separated from the closest server by a distance twice as long as the hinterland. The interior servers are paired or isolated.

As an example, consider the game $H(6, [0, 1])$. Let the length of the hinterland be x, and without loss of generality let the servers be ordered such that $x_1 < x_2 < \ldots < x_6$. From the above characterization of equilibria, it follows that in every equilibrium s_1 and s_2 are paired at $x_1 = x_2 = x$, s_3 is located at $x_3 = 3x$, s_4 is located at $x_4 = 1 - 3x$ and s_5 and s_6 are paired at $x_5 = x_6 = 1 - x$. The distance between s_3 and s_4 is at least ε and at most $2x$ (by condition (El2)).

That is, the above described configuration is an equilibrium for every x such that $\varepsilon \leq (1 - 3x) - 3x \leq 2x$, or rather $1/8 \leq x < 1/6$. (See Fig. 9.2.)

9.3 Hotelling Game on the Line with Link Failures

Let us now consider the game with environmental failures. For concreteness, let us assume that the only possible failure is a disconnection of the line at point f, chosen uniformly from $[0, 1]$, which severs the line into two separate markets, and forces some of the clients (specifically, those disconnected from their originally chosen server) to change their server selection. For simplicity, assume at most one disconnection may occur, with constant probability $0 < r < 1$, at a location $f \in [0, 1]$ chosen uniformly at random. We call this game the *Line Failure Hotelling game* and denote it as $H_{\mathrm{LF}}(n, r, [0, 1])$. The new payoff function is denoted as p_{LF} and becomes the expected profit under these assumptions (i.e., the payoff of player s_i is $1 - r$ times its payoff in the fault-free case plus r times its expected payoff in case a disconnection occurred at point $0 < f < 1$).

Let us begin by considering this game with only *one server*. Unlike the basic Hotelling game, in which the location of a single server is inconsequential, in this setting the optimal location of a single server is at the center of the line, $x_1 = 1/2$, as we show next. Let $x_1 \in [0, 1]$ be the location of the server s_1. If no failure occurs, the payoff is 1. If the line is disconnected at $0 < f < x_1$, then the payoff is $1 - f$. If the line is disconnected at $x_1 < f < 1$, then the payoff is f. It follows that the payoff of s_1 is

$$p_{\mathrm{LF}}(s_1) = \mathbb{E}[p(s_1)] = (1 - r) \cdot 1 + r \cdot \left[\int_0^{x_1} (1 - f)df + \int_{x_1}^1 f df \right]$$

$$= 1 - \frac{r}{2} + rx_1 - rx_1^2 .$$

That is, $p_{\mathrm{LF}}(s_1)$ is a function of x_1 that attains its maximum at $x_1 = 1/2$. Hence the only equilibrium is when the server is at the center of the market.

We next consider the *two server* variant. Without loss of generality let $x_1 < x_2$, i.e., s_1 is located to the left of s_2. Observe that the payoff of s_1 is $p_{\mathrm{LF}}(s_1) = L_{\mathrm{LF}}(s_1) + R_{\mathrm{LF}}(s_1)$, where $L_{\mathrm{LF}}(s_1)$ is the length of its hinterland, and $R_{\mathrm{LF}}(s_1)$ is the length of its half-market on the right.

- **Calculating** $\mathbf{L_{LF}(s_1)}$: if the line is disconnected at location $0 < f < x_1$, then the length of s_1's hinterland is $L(s_1) = x_1 - f$. Otherwise, $L(s_1) = x_1$. The expected length is therefore

$$L_{\mathrm{LF}}(s_1) = \mathbb{E}[L(s_1)] = (1 - r) \cdot x_1 + r \cdot \left[\int_0^{x_1} (x_1 - f)df + \int_{x_1}^1 x_1 df \right]$$

$$= x_1 - \frac{r \cdot x_1^2}{2} .$$

- **Calculating** $\mathbf{R_{LF}(s_1)}$: if no failure occurs, or if the line is disconnected at location $0 < f < x_1$ or $x_2 < f < 1$, then the length of the half-market is $R(s_1) = (x_2 - x_1)/2$. If, on the other hand, an edge is disconnected at $x_1 < f < x_2$, then $R(s_1) = f - x_1$. Therefore,

$$R_{\mathrm{LF}}(s_1) = \mathbb{E}[R(s_1)] = (1 - r) \cdot \frac{x_2 - x_1}{2} + r \cdot \frac{\int_{x_1}^{x_2} (f - x_1)df}{x_2 - x_1} = \frac{x_2 - x_1}{2} .$$

It follows that the payoff of s_1 is

$$p_{\mathrm{LF}}(s_1) = L_{\mathrm{LF}}(s_1) + R_{\mathrm{LF}}(s_1) = x_1 - \frac{r \cdot x_1^2}{2} + \frac{x_2 - x_1}{2} = \frac{x_2 + x_1}{2} - \frac{r \cdot x_1^2}{2} .$$

This function attains its maximum at $x_1 = 1/(2r)$. That is, as long as s_1 remains on the left of s_2, s_1 would move to $1/(2r)$. If $1/2 \leq x_2 < 1/(2r)$, then s_1's best response would be pairing with s_2.

Note that $1/(2r) > 1/2$ for every $0 < r < 1$ and thus if $x_1 = 1/(2r)$ and $x_2 > 1/(2r)$, then s_2 would prefer to move to the left of s_1. It follows that $x_2 < 1/(2r)$. By symmetry, it also holds that $x_1 > 1 - 1/(2r)$. Hence, condition (EL1) holds, and as in the basic Hotelling game, the only equilibrium is when both servers are paired at the center of the line, i.e., $x_1 = x_2 = 1/2$.

Consequently, for the general game, with $n \geq 3$ servers, we have the following.

Observation 1 *In the Hotelling game $H_{\mathrm{LF}}(n, r, [0, 1])$, the following holds:*

1. *A peripheral server located at distance x from the boundary has a hinterland with an expected length of $x - rx^2/2$.*
2. *Two neighboring servers at distance x gain a half-market of expected length $x/2$ each in the direction of the other (as in the basic Hotelling game).*

3. *Each peripheral server would increase his hinterland up to $1/(2r)$. That is, condition (EL1) holds unless the neighbor of a peripheral server is at a distance of more than $1/(2r)$ from the boundary.*

In light of the observation above, if the game is played with *three servers*, then no equilibrium exists. To see why, observe that if the interior server s_2 is located between $1 - 1/(2r)$ and $1/(2r)$, then the peripheral servers s_1 and s_3 would pair with s_2 on both sides leaving it with 0 payoff, and thus s_2 would move. If, on the other hand, s_2 is located at $x_2 > 1/(2r)$, then s_1 would locate at $1/(2r)$ and s_3 would pair with s_2. But then s_3 could improve its payoff and would thus move. Due to symmetry, a similar argument holds if we suppose s_2 is located at $x_2 < 1 - 1/(2r)$.

Next, consider the *four server* game $H_{\mathrm{LF}}(4, r, [0, 1])$. Let us consider the strategy profile where the location $0 < x < 1$ satisfies that when s_1 and s_2 are paired at x and s_3 and s_4 are paired at $1 - x$ all four servers receive the same expected payoff. The detailed calculation is deferred to the full paper [1], but this occurs when $x = (2 - \sqrt{4 - r})/r$, and by Observation 1 this is a Nash equilibrium.

Moreover, this Nash equilibrium is unique due to the following considerations. First, by Observation 1, the peripheral servers must be paired with their neighbors. Second, the peripheral servers must have hinterlands of the same length, otherwise one would take the hinterland of the other. Third, every two paired servers must have the same expected payoff, otherwise one would take the half-market of the other. It follows that the configuration above is the only Nash equilibrium.

We next consider the general n-server game $H_{\mathrm{LF}}(n, r, [0, 1])$. By Observation 1, for any given configuration of servers, adding link failures to the game only affects the expected payoff gained from the hinterlands. Namely, a hinterland of length x shrinks by $rx^2/2$ in expectation, while every other half-market retains its length in expectation. This leads to the following theorem. (Hereafter, proofs are deferred to the full paper, see [1].) Let $a = rx_1^2/2$ and $b = 1 - rx_n^2/2$, hence

$$[a, b] = \left[\frac{r \cdot x_1^2}{2}, 1 - \frac{r \cdot (1 - x_n)^2}{2} \right].$$

Theorem 2 *Let $0 \leq x_1 \leq x_2 \leq \ldots \leq x_n \leq 1$. The configuration $(x_1, x_2, \ldots, x_n)$ is a Nash equilibrium of the game $H_{\mathrm{LF}}(n, r, [0, 1])$ if and only if it is a Nash equilibrium of the game $H(n, [a, b])$.*

Theorem 2, in conjunction with conditions (El1) and (EL2) for an equilibrium of the basic Hotelling game, yields the following conditions for an equilibrium of the Hotelling game with line disconnections.

Corollary 1 *The following conditions are sufficient and necessary for an equilibrium of the Hotelling game on the line with line disconnections.*

(1) The peripheral servers are paired, and are located at an identical distance from the boundary, x.
(2) No interior server's whole market is smaller than $x - r \cdot x^2/2$.
(3) No interior server's half market is larger than $x - r \cdot x^2/2$.

To summarize, this section established the following theorem.

Theorem 3 *In the n-server Hotelling game $H_{LF}(n, r, [0, 1])$*

(i) *For $n = 1$, a Nash equilibrium exists with $x = 1/2$.*
(ii) *For $n = 2$, there exists an equilibrium with $x_1 = x_2 = 1/2$.*
(iii) *For $n = 3$, no equilibrium exists.*
(iv) *For $n = 4$, there exists an equilibrium with $x_1 = x_2 = (2 - \sqrt{4-r})/r$, $x_3 = x_4 = 1 - (2 - \sqrt{4-r})/r$.*
(v) *For $n = 5$, there exists an equilibrium with $x_1 = x_2 = (3 - \sqrt{9-4r})/(2r)$, $x_4 = x_5 = 1 - (3 - \sqrt{9-4r})/(2r)$ and $x_3 = 1/2$.*
(vi) *For $n \geq 6$, there exist infinitely many equilibria, characterized as follows: peripheral servers are paired with their neighbor and have identical hinterlands of length x. Each peripheral pair is separated from the closest server by a distance twice as long as $x - r \cdot x^2/2$. The interior servers are paired or isolated such that no server's whole market is smaller than $x - r \cdot x^2/2$, and no server's half-market is larger than $x - r \cdot x^2/2$.*

9.4   Hotelling Game on the Line with Server Crashes

Let us now consider a different failure setting, where the environment is resilient, but the servers might crash. For concreteness, let us assume that each server might fail with probability $0 < r < 1$ independently of the others. Once a server has failed, the clients who chose this server originally must change their server selection. We call this game the *Player Failure Hotelling game* and denote it as $H_{PF}(n, r, [0, 1])$. The new payoff function is again the expected profit under these assumptions and is denoted as p_{PF}.

For $n = 1$, it is clear that the server's expected payoff is $p_{PF}(s_1) = 1 - r$ wherever its location is. For $n = 2$, it is easy to see that server crashes have no impact; the game is equivalent to the basic Hotelling game and the only Nash equilibrium is when the servers are paired at the center.

Let us now analyze the case where there are *three servers* on the line. In every equilibrium, the peripheral servers are paired with the interior server on both sides, and the interior server is located at $1/2$. This is because when pairing with the interior server, each peripheral server is closest to its neighbor regardless of which servers crash. Moreover, if they are not located at the center, then one peripheral server could improve its payoff by taking the hinterland of the other. Note that, in this state, the expected payoff of the peripheral server s_1 is

$$p_{PF}(s_1) = \mathbb{E}[p(s_1)] = (1 - r)\left(\frac{1}{2} + \frac{1}{2} \cdot r^2\right),$$

since it always gets its hinterland (provided it did not fail), and it gets the remainder of the line only if both other servers have failed. By a similar case analysis, the expected payoff of the interior server s_2 is

$$p_{PF}(s_2) = \mathbb{E}[p(s_2)] = (1-r)\left(0 + \frac{1}{2} \cdot r(1-r) \cdot 2 + 1 \cdot r^2\right),$$

since (provided it did not fail) it gets nothing if both other servers do not fail, it gets half the line if one of the other servers has failed, and it gets the entire line if both other servers have failed. Note that s_2 can move to take s_1's hinterland, hence this state would be a Nash equilibrium only if $p_{PF}(s_1) = p_{PF}(s_2)$, i.e.,

$$(1-r)\left(\frac{1}{2} + \frac{1}{2} \cdot r^2\right) = (1-r)\left(0 + \frac{1}{2} \cdot r(1-r) \cdot 2 + 1 \cdot r^2\right),$$

which yields $r^2 - 2r + 1 = 0$, whose only solution is $r = 1$, i.e., all servers crash in every game, which is obviously an equilibrium, but not an interesting one. This proves that no equilibrium exists when there are three servers in the game.

We expand this logic to make a general claim about the game.

Theorem 4 *For every $n \geq 3$, the Player Failure Hotelling game $H_{PF}(n, r, [0, 1])$ has no Nash equilibrium in pure strategies.*

Proof Sketch Observe that in this game, interior servers must "think" like peripheral servers, i.e., each server s_i tends to move away from a boundary and towards one of its neighbors. This holds because each interior server s_i has some probability of becoming a peripheral server, in which case s_i would profit by increasing the hinterland. However, s_i is more likely to become the peripheral server of the boundary closest to it (separated by fewer servers), and thus would do better in expectation by moving away from this boundary and towards the center. It follows that all servers would converge towards the center, which clearly never results in an equilibrium.

9.5 Conclusion

In this paper we considered two fault tolerant variants of the Hotelling game: link failures and player faults. On the one hand, we have shown that the game is resilient to link failures in some sense—each equilibrium is related to an equilibrium of the no-faults Hotelling game by rescaling the interval along with player positions. On the other hand, we have shown that the game is vulnerable to player failures. No equilibrium exists because players tend to converge towards the center.

There are many possible future directions for this research. A large number of variants of the Hotelling game have been studied and each would be interesting to consider in a faulty setting, such as: the Hotelling game on graphs, on the plane or over $\mathbb{R}^n$, the Hotelling game with sequential entry, and so on. Another interesting direction would be to try other fault models. Some examples are models where the number of faulty players is bounded, where faulty players remain in the game but act unexpectedly (or "Byzantinely"), or where faults are injected adversarially rather than at random.

References

1. Avin, C., Cohen, A., Lotker, Z., Peleg, D.: Fault-tolerant hotelling games. CoRR abs/1801.04669 (2017). http://arxiv.org/abs/1801.04669
2. Chechik, S., Peleg, D.: Robust fault tolerant uncapacitated facility location. Theor. Comput. Sci. **543**, 9–23 (2014)
3. Chechik, S., Peleg, D.: The fault-tolerant capacitated k-center problem. Theor. Comput. Sci. **566**, 12–25 (2015)
4. De Palma, A., Ginsburgh, V., Thisse, J.F.: On existence of location equilibria in the 3-firm hotelling problem. J. Ind. Econ. **36**, 245–252 (1987)
5. Eaton, B.C., Lipsey, R.G.: The principle of minimum differentiation reconsidered: some new developments in the theory of spatial competition. Rev. Econ. Stud. **42**(1), 27–49 (1975)
6. Eiselt, H.A.: Equilibria in competitive location models. In: Foundations of Location Analysis, pp. 139–162. Springer, Berlin (2011)
7. Eiselt, H.A., Laporte, G., Thisse, J.F.: Competitive location models: a framework and bibliography. Transp. Sci. **27**(1), 44–54 (1993)
8. Gradwohl, R., Reingold, O.: Fault tolerance in large games. Games Econ. Behav. **86**, 438–457 (2014)
9. Hotelling, H.: Stability in competition. Econ. J. **39**(153), 41–57 (1929)
10. Khuller, S., Pless, R., Sussmann, Y.J.: Fault tolerant k-center problems. Theor. Comput. Sci. **242**(1–2), 237–245 (2000)
11. Snyder, L.V., Atan, Z., Peng, P., Rong, Y., Schmitt, A.J., Sinsoysal, B.: Or/ms models for supply chain disruptions: a review. IIE Trans. **48**(2), 89–109 (2016)
12. Sviridenko, M.: An improved approximation algorithm for the metric uncapacitated facility location problem. In: International Conference on Integer Programming and Combinatorial Optimization, pp. 240–257. Springer, Berlin (2002)
13. Swamy, C., Shmoys, D.B.: Fault-tolerant facility location. ACM Trans. Algorithms **4**(4), 51 (2008)
14. Wang, X., Ouyang, Y.: A continuum approximation approach to competitive facility location design under facility disruption risks. Transp. Res. B Methodol. **50**, 90–103 (2013)
15. Zhang, Y., Snyder, L.V., Ralphs, T.K., Xue, Z.: The competitive facility location problem under disruption risks. Transp. Res. E Logist. Transp. Rev. **93**, 453–473 (2016)

Chapter 10
A Social Choice Theoretic Approach for Analyzing User Behavior in Online Streaming Mobile Applications

Neetu Raveendran, Kaigui Bian, Lingyang Song, and Zhu Han

10.1 Introduction

Social choice theory is a framework that analyzes how a group of individuals arrives at a collective decision, called the *social choice*, from among a set of individual preferences or opinions. It is assumed that each individual has a preference relation for a given set of alternatives [9]. Social choice deals with combining these preference relations into a single preference relation that represents the preference of the whole group. The set of individual preference relations is called a *preference profile* [4]. Decision making scenarios in which social choice theory finds application are ubiquitous; electing a leader from among a set of political candidates in a democracy, choosing a chairman from among the union members, selecting a winner from among the contestants in a reality television show, and so on.

Social choice theory is prevalent in many research areas. Geist and Peters [6] relies on computer-aided methods like SAT solvers to solve NP-complete problems in social choice. A few research directions in the inter-disciplinary field known as *computational social choice*, which combines social choice theory and computer science, are discussed in [4]. Anshelevich and Postl [2] studies a scenario where

N. Raveendran (✉)
Department of Electrical and Computer Engineering, University of Houston, Houston, TX, USA
e-mail: nraveendran@uh.edu

K. Bian · L. Song
School of Electronics Engineering and Computer Science, Peking University, Beijing, China
e-mail: bkg@pku.edu.cn; lingyang.song@pku.edu.cn

Z. Han
Department of Electrical and Computer Engineering, University of Houston, Houston, TX, USA

Department of Computer Science and Engineering, Kyung Hee University, Seoul, South Korea
e-mail: zhan2@uh.edu

© Springer Nature Switzerland AG 2019
J. B. Song et al. (eds.), *Game Theory for Networking Applications*,
EAI/Springer Innovations in Communication and Computing,
https://doi.org/10.1007/978-3-319-93058-9_10

each individual has a cost associated with each alternative, and also, evaluates the effectiveness of randomized social choice algorithms. A novel decision making scheme known as *Maximal Recursive (MR)*, which has better efficiency and computability compared to an existing alternative is proposed in [3].

The current era of enormous social network and live streaming traffic shows great potential for a social choice study, in order to better learn user behavior. Social choice theory is used to study decision making in choices of online services by users in [7]. A user preference matrix is created firstly, according to each user's preferences. Subsequently, using this matrix, 0–1 *Integer Linear Programming (ILP)* is performed on *Kemeny's social choice functions*, which is discussed in detail later. As a result, an optimal ranking of the online services, along the lines of the common social preference is obtained.

Even though social choice analysis is performed on online services using Kemeny's functions in [7], an evaluation of online live streaming applications, considering other important aspects of social choice has not been performed previously, according to the authors' knowledge. Therefore, in this paper, the concepts of social choice theory are employed in exploring the characteristics of the online live streaming paradigm. Accordingly, we utilize the social choice model to analyze a real-time dataset from an online live streaming application and draw insights on the user behavior. We also formulate our considered scenario as an ILP optimization problem similar to [7], and validate the results from the social choice study with the results obtained by simulating the ILP optimization. The results achieved through our analysis depict the effectiveness of the model in evaluating the group preferences of users in online live streaming applications, and also imply the feasibility of the model in other similar research.

We organize the rest of this paper as follows. In Sect. 10.2, we firstly introduce the system model, and then, formulate the considered social choice model as an ILP. We discuss the results of the analysis performed on the procured live streaming dataset in Sect. 10.3. Here, we explore the application of social choice theory on the dataset, and discuss a few insights in Sect. 10.3.1. Thereafter, we validate the results obtained from this social choice analysis through the formulated ILP in Sect. 10.3.2. Finally, based on our analysis, we draw conclusions in Sect. 10.4.

10.2   System Model and Problem Formulation

We consider an online live streaming scenario, where performers can transmit their performances through a live multimedia streaming platform, and users can stream and watch these performances live on their mobile devices or computers. Accordingly, we consider a scenario with n performers and m users. Each of the m users have their individual preferences for each of the n performers, according to their interests in the streamed content. The users can also send messages and virtual gifts to their preferred performers.

In order to utilize the concepts of social choice theory to better analyze the considered scenario, we need to combine the individual preference relations of the users into a single preference relation representing all the users. To this end, we represent a preference relation of user u_i, where $i \in \{1, 2, \ldots, m\}$, over the set of alternatives, which is the set of performers $\{p_1, p_2, \ldots, p_n\}$, as

$$p_j \succ_i p_k, \tag{10.1}$$

which denotes that user u_i prefers performer p_j to performer p_k.

Next, we define three matrices, $\mathbf{R}, \mathbf{C}$, and $\mathbf{E}$, which reflect the user preferences on performers, and an $\mathbf{S}$ matrix reflecting the group preferences on performers, as shown below, similar to [7].

- $\mathbf{R} = [r_{ij}]_{m*n}$, where $r_{ij} \in \{1, 2, \ldots, n\}, \forall i \in \{1, 2, \ldots, m\}$, and $\forall j \in \{1, 2, \ldots, n\}$ consists of scores assigned by the users to the performers according to their preferences. Here, each of the m rows of the matrix, representing the m users consists of n columns, containing the scores assigned to each of the n performers. The higher the score assigned, the more preferred the performer; for example, if $r_{11} > r_{12}$, then $p_1 \succ_1 p_2$, i.e., user u_1 prefers performer p_1 to performer p_2.
- $\mathbf{C} = [c_{jk}]_{n*n}, \forall j, k \in \{1, 2, \ldots, n\}$ consists of the count of users preferring each of the n performers to each of the other $n - 1$ performers, which can be expressed as

$$c_{jk} = \sum_{i=1}^{m} I(p_j \succ_i p_k), \tag{10.2}$$

where I is an indicator function, which can be defined as

$$I(p_j \succ_i p_k) = \begin{cases} 1, & p_j \succ_i p_k, \\ 0, & p_j \prec_i p_k. \end{cases} \tag{10.3}$$

- $\mathbf{E} = [e_{jk}]_{n*n}, \forall j, k \in \{1, 2, \ldots, n\}$ denotes the ratio of the difference between the number of users who prefer performer p_j to performer p_k and the number of users who prefer performer p_k to performer p_j, to the number of users, which can be defined as

$$e_{jk} = \frac{(c_{jk} - c_{kj})}{m}. \tag{10.4}$$

- $\mathbf{S} = [s_{jk}]_{n*n}, \forall j, k \in \{1, 2, \ldots, n\}$ denotes the group preferences on performers, which is the social choice in our scenario, and can be defined as

$$s_{jk} = \begin{cases} 1, & p_j \succ p_k, \\ 0, & p_j \prec p_k. \end{cases} \tag{10.5}$$

Note that the preferences here are denoted by $\succ$ and $\prec$ rather than $\succ_i$ and $\prec_i$, as these are the preferences of the whole group.

As discussed before, social choice is a model to combine individual preferences into a single preference representing a whole group. In such a model, a function that maps the preference profiles of the individuals to the most preferred alternative of the group is called a *social choice function*. There are different social choice functions like the *majority rule*, the *Borda count* [13], and so on. Kemeny's functions is a category of social choice functions for group decision making like voting, and maximizes the similarity between the preference of the group and the individual preferences [7]. Kemeny's function assigns a value for each ranking to indicate this similarity, and the social choice is the ranking with the maximum value.

Computation of Kemeny's functions have been mentioned to be NP-hard in [1]. Therefore, [7] determines the optimal social ranking of online services for Kemeny's functions using a 0–1 ILP. ILP problems are constraint optimization problems, where the optimization of linear functions with binary integer variables is the objective [8].

The 0–1 ILP to obtain the Kemeny's functions as in [7] can be formulated for our scenario as

$$f_{kem} = 2 \max \sum_{j,k=1}^{n} e_{jk} s_{jk},$$

$$\text{s.t.} \begin{cases} s_{jk} \in \{0, 1\}, \\ 0 \le s_{hj} + s_{jk} - s_{hk} \le 1, \end{cases}$$

(10.6)

$\forall h < j < k$, where f_{kem} gives the Kemeny's function value.

Here, the first constraint makes sure that the s_{jk} values are in the range given by (10.5). The second constraint assures that the preferences are transitive, i.e., if performer p_h is preferred to performer p_j, and performer p_j is preferred to performer p_k, then, performer p_h is preferred to performer p_k, $\forall h < j < k$. The above optimization yields the social choice matrix values, $\mathbf{S} = [s_{jk}]_{n*n}, \forall j, k \in \{1, 2, \ldots, n\}$, which can be considered as an indicator function taking a value equal to 1, if the number of users preferring performer p_j to performer p_k is more than the number of users preferring performer p_k to performer p_j, and taking a value of 0, otherwise. In short, the values of the social choice matrix $\mathbf{S}$, obtained through the above optimization, provides the group preference relation for the m users on the n performers.

10.3   Analysis and Discussion

In this section, firstly, we show the results from the analysis of a real-time dataset acquired from an online live streaming platform, in Sect. 10.3.1. Thereafter, in Sect. 10.3.2, we utilize the 0–1 ILP formulation as shown before, to validate the results from the real-time data analysis.

10.3.1   Analysis of Real-Time Dataset

Online live streaming has been gaining tremendous popularity in China since 2015, with more than 325 million total viewers [11, 12]. Hence, a real-time dataset from such a set of users would be a microcosm of the overall online live streaming audience. Accordingly, we were able to obtain a dataset from a Chinese live streaming mobile application named *Bangyang*. We performed an analysis of the dataset obtained for the months of May 2015–March 2016, with a monthly average of around 1344 users and 53,386 chat messages, with the help of SQL. The chat messages were sent by the users to the performers; the more preferred the performer, the higher the number of chat messages sent by the user. Therefore, we made use of the number of chat messages as a parameter to understand the preference relations of the users, and those of the whole group.

As the platform went through around thousands of users and around 50,000 of chat messages per month, we needed a subset of this traffic, in order to better understand the preference profiles of the users and the social choice of the whole group. For this purpose, we short-listed ten users who contributed the largest shares of chat message traffic to the performers. Then, we analyzed the top five performers to whom each of these users sent the most number of messages, which forms the preference profiles of these ten users. We also analyzed the data to determine the top three performers receiving the most number of chat messages overall.

Figure 10.1 demonstrates the results of this analysis using the data for May 2015, a month which witnessed 77,839 chat messages, 1934 performers, and 2167 users. Due to privacy reasons, the performer IDs are denoted as P1, P2, and P3 here. We can observe that the first preference for user 1 is P1, second preference is P2, but the third, fourth, and fifth preferences are not among the top three performers, and hence, we have marked those as irrelevant. Similarly, we can also observe the preference lists of users 2–10 in the figure. Additionally, the topmost row denotes the overall top three performers, which is indeed the preference profile of the group, i.e., the social choice.

Top Performers		1st	2nd	3rd	4th	5th
		P1	P2	P3		
Preference Profile	User 1	P1	P2	Irrelevant	Irrelevant	Irrelevant
	User 2	P1	Irrelevant	Irrelevant	Irrelevant	Irrelevant
	User 3	P2	Irrelevant	Irrelevant	Irrelevant	Irrelevant
	User 4	Irrelevant	P1	Irrelevant	Irrelevant	Irrelevant
	User 5	Irrelevant	Irrelevant	P2	P1	Irrelevant
	User 6	P1	Irrelevant	Irrelevant	Irrelevant	Irrelevant
	User 7	P1	P2	Irrelevant	Irrelevant	Irrelevant
	User 8	Irrelevant	Irrelevant	Irrelevant	P1	Irrelevant
	User 9	P1	Irrelevant	P2	Irrelevant	P3
	User 10	P3	P1	Irrelevant	Irrelevant	Irrelevant

Fig. 10.1 Preference profiles of users and social choice

We can clearly observe from Fig. 10.1 that out of the top ten users, 90% had P1, 50% had P2, and 20% had P3 in the top five preferences. This reinforces the group preference profile, i.e., the *social choice* obtained through the data analysis, as P1 ≻ P2 ≻ P3.

Such a function which maps the preference profiles of the users to a preference relation for the whole group, as seen from the above analysis, is called a *social welfare function*, which is the key element in one of the most important results of social choice theory. This concept, known as the *impossibility theorem* put forward by Kenneth Arrow [5, 9, 14], can be applied in our scenario after understanding the *independence of irrelevant alternatives* property, which is demonstrated using our analysis results in Fig. 10.2.

According to this property, when the alternatives which are irrelevant to the obtained group preference relation are removed from the set of alternatives, and then a group preference relation is reevaluated, it should not be different from the initially obtained group preference relation. We can observe that from Fig. 10.1, if we remove the performers marked as *irrelevant*, we obtain the preference profiles as shown in Fig. 10.2. Now, if we reevaluate the group preference relation, we still get P1 ≻ P2 ≻ P3, which is the group preference initially obtained. Hence, our considered scenario satisfies this property.

Now, we consider the relevance of the much celebrated Arrow's theorem in our scenario. According to this theorem, when there are three or more alternatives, a social welfare function will either not select *unanimously* from a set of individual preferences, or will be similar to a *dictatorship*, where one particular individual's preference is selected to be the group preference, or will not hold the *independence of irrelevant alternatives* property [9, 10].

Top Performers		1st	2nd	3rd
		P1	P2	P3
Preference Profile	User 1	P1	P2	
	User 2	P1		
	User 3	P2		
	User 4	P1		
	User 5	P2	P1	
	User 6	P1		
	User 7	P1	P2	
	User 8	P1		
	User 9	P1	P2	P3
	User 10	P3	P1	

Fig. 10.2 Independence of irrelevant alternatives

From Figs. 10.1 and 10.2, we can observe that even though our scenario, which consists of more than three alternatives, satisfies the *independence of irrelevant alternatives* and the *non-dictatorship* properties, the group preference relation (P1 $\succ$ P2 $\succ$ P3) is not chosen *unanimously* by all the top ten users in our dataset, i.e., P1 $\succ$ P2 $\succ$ P3 is not the preference relation for all the ten users. This clearly demonstrates the relevance of Arrow's theorem in our scenario.

10.3.2 Validation Using ILP

Next, we utilize the 0–1 ILP formulation to obtain the Kemeny's functions as shown in (10.6), to validate the data analysis results we discussed. To this end, we implement MATLAB simulations to obtain the values in the group preference matrix, $\mathbf{S} = \left[s_{jk} \right]_{n*n}, \forall j, k \in \{1, 2, \ldots, n\}$.

Figure 10.3 shows the values used in the $\mathbf{R}$ matrix, which are the scores assigned by the users to the performers according to their individual preferences. These are the scores determined by considering the preference profile of each of the top ten users, as shown in Fig. 10.1. For example, for user 9, out of the top five preferences, P1 comes first, P2 comes third, and P3 comes fifth, and hence, the scores for performers 1, 2, and 3 are 5, 3, and 1, respectively. Similarly, we can comprehend the scores assigned by the other users to the performers.

Fig. 10.3 **R** matrix with user scores for performers

	Performer 1	Performer 2	Performer 3
User 1	5	4	0
User 2	5	0	0
User 3	0	5	0
User 4	4	0	0
User 5	2	3	0
User 6	5	0	0
User 7	5	4	0
User 8	2	0	0
User 9	5	3	1
User 10	4	0	5

Fig. 10.4 **S** matrix obtained using ILP demonstrating the social choice

	Performer 1	Performer 2	Performer 3
Performer 1	0	1	1
Performer 2	0	0	1
Performer 3	0	0	0

The values in the **S** matrix obtained by simulating the 0–1 ILP in (10.6) are shown in Fig. 10.4. The value of 1 in the first row and second column indicates that in the group preference relation, performer 1 is preferred to performer 2. Similarly, the remaining values of 1 denote that performer 1 is preferred to performer 3, and performer 2 is preferred to performer 3 in the group preference relation, resulting in P1 ≻ P2 ≻ P3 as the *social choice*, which is exactly the same as observed in the analysis results in Fig. 10.1. Thus, the results from the ILP formulation substantiate the data analysis results in Sect. 10.3.1.

10.4 Conclusions

This paper makes use of social choice theory as a model to aggregate individual preferences in a group to a single preference relation for the entire group. It is clearly demonstrated in the analysis that social choice is an effective framework for exploring the behavior of online live streaming applications, which have immense popularity among current internet users. We observe from the *Bangyang* dataset analysis results that social choice distinctly represents the aggregated performer preferences of the considered group of users. We also underline the prominence of concepts like the independence of irrelevant alternatives and Arrow's theorem in the considered scenario. Furthermore, we validate the results obtained through the social choice model analysis using an ILP model simulation of our scenario. In this paper, as the scope of evaluation is a real-time dataset obtained from a live streaming platform with huge traffic, it is conspicuous that the social choice model can be employed to further research similar scenarios.

Acknowledgements This research is partially supported by US NSF CNS-1717454, CNS-1731424, CNS-1702850, CNS-1646607, ECCS-1547201, China NSFC 61632017.

References

1. Ali, A., Meila, M.: Experiments with Kemeny ranking: what works when? Math. Soc. Sci. **64**(1), 28–40 (2012)
2. Anshelevich, E., Postl, J.: Randomized social choice functions under metric preferences. In: Proceedings of the 25th International Joint Conference on Artificial Intelligence (IJCAI), pp. 46–59 (2016)
3. Aziz, H., Postl, J.: Maximal recursive rule: a new social decision scheme. In: Proceedings of the 23rd International Joint Conference on Artificial Intelligence (IJCAI), pp. 34–40 (2013)
4. Aziz, H., Brandt, F., Elkind, E., Skowron, P.: Computational social choice: the first ten years and beyond. In: Steffen, B., Woeginger, G. (eds.) Computer Science Today. Lecture Notes in Computer Science (LNCS), vol. 10000. Springer, New York (2017)
5. Feiwel, G.R.: Arrow and the Ascent of Modern Economic Theory. New York University Press, New York (1987)
6. Geist, C., Peters, D.: Computer-aided methods for social choice theory. In: Endriss, U. (ed.) Trends in Computational Social Choice, Chap. 13, AI Access (2017)
7. Li, W., Fu, X., Huang, Q., Liu, L.: Evaluating on online services based on social choice theory. In: 2016 Chinese Control and Decision Conference, pp. 6512–6517 (2016)
8. Marinescu, R., Dechter, R.: Best-first AND/OR search for 0/1 integer programming. In: Integration of AI and OR Techniques in Constraint Programming for Combinatorial Optimization Problems (CPAIOR), pp. 171–185 (2007)
9. Maschler, M., Solan, E., Zamir, S.: Game Theory. Cambridge University Press, Cambridge (2013)
10. Sen, A.: Social choice theory: a re-examination. Econometrica **45**(1), 53–89 (1977)
11. Livestreaming in China: Everything you need to know. https://econsultancy.com/blog/69137-livestreaming-in-china-everything-you-need-to-know
12. Live Streaming in China: The Top 5 Platforms. https://chozan.co/2017/05/02/china-top-5-live-streaming-platforms/
13. Social Choice Theory - Stanford University. https://web.stanford.edu/class/symbsys150/social-choice-theory-5-8.html
14. The birth of social choice theory from the spirit of mathematical logic: Arrow's theorem in the framework of model theory. http://www100.uni-graz.at/vwlwww/forschung/RePEc/wpaper/2016-04.pdf

Chapter 11
Social Coordination and Network Formation with Heterogeneous Constraints

Qingchao Zeng

11.1 Introduction

In many economic and social situations, people can benefit from choosing the same action or following the same standard. Coordination games with multiple Nash-Equilibria can explain the internal mechanisms in those situations. Many literature have explored the mechanisms by which conventions become established under fixed interaction structures in coordination games. And the most important result is that under myopic best response rules risk dominate conventions will be the long-run equilibria.

In this paper, we set up a model where players decide which action to choose in a coordination game and who they connect with. We aim to understand the mechanism in a setting where agents are divided into two groups via different constraints of links. We encompass two groups of agents called N_1 and N_2. The agents in the N_1 support a smaller limit number of links than the players in the N_2 for the reason of less resources, which is different with the scenario where the overall population shares the same limit number of links à la Staudigl and Weidenholzer [23]. This linking constraints are reasonable since the number of population with whom one agent links is fairly small compared with the overall population in many situations. Moreover, giving a example of Weibo (a social network application which is similar with Twitter), general users can only follow no more than 2000 others, where the upper bound of users that VIP users can follow is 20,000, where the daily average number of active users is more than 1 billion. For instance, such constrained interactions will arise in many circumstances where the time agent can spend on social network is limited and different, and there are increasing marginal costs where agents' affordability is different.

Q. Zeng (✉)
School of Economics and Management, Beihang University, Beijing, People's Republic of China

© Springer Nature Switzerland AG 2019   153
J. B. Song et al. (eds.), *Game Theory for Networking Applications*,
EAI/Springer Innovations in Communication and Computing,
https://doi.org/10.1007/978-3-319-93058-9_11

In our model, we consider a model of one side network formation where each agent can decide on the set of players whom to connect with costly *à* la Staudigl and Weidenholzer [23]. The agents in the group N_1(and N_2) may connect with other agents both in group N_1 and N_2, and those two connected players play a 2×2 coordination games with each other. In each period, players in the two groups select an action to play in all of her connections and the choice of linking sets to maximize their respected payoffs under the myopic best response rules.

The structure of this paper is as follows. Section 11.2 reviews the related literature of social coordination and network formation. Section 11.3 describes the model and introduces the significant methods used. Section 11.4 considers the constrained interactions and indicates the main results we find. Section 11.5 presents the conclusion on our discussion.

11.2   Related Literature

There are lots of previous works related to interactions within the same population, for global interaction models see, e.g., Kandori et al. [22], Kandori and Rob [21], Young [25] and for local interaction models see, e.g., Blume [7, 8], Ellison [10, 11] and Alós-Ferrer and Weidenholzer [2]. A branch of present literature related to social coordination and network formation considers scenario where the agent has to decide on both the action choice and the set of interaction partners. In Goyal and Vega-Redondo [16], agents unilaterally decide on whom to connect with costly. And the payoff of coordination games is received by both sides of the link whereas in our model the payoff is only received by the active side. They find that when the linking costs are relatively low, the risk-dominate convention is selected whereas the payoff-dominate convention emerges when the costs of maintaining a link are relatively high. In Hojman and Szeidl [18] , they set up a model where agents only have to pay for their out-degree links but can receive payoffs from all path-connected partners. Weidenholzer [24] presents a local interaction model where agents only can interact with a small subset of the population rather than the overall population. They show that on the scenario where myopic best response rules is used and the interaction structure is fixed, the risk-dominate convention will emerge in long run. In Jackson and Watts [19], they propose a setup in which forming a link needs the agreement of both sides of the interactions. They find that the selection of conventions depends on the level of linking cost: the risk dominate convention is selected for low linking costs whereas when the linking costs are high, both risk dominate convention and payoff dominate convention are selected. Nevertheless, they do not find any evidence indicating that constrained interactions have a significant impact on the selection of conventions. Staudigl and Weidenholzer [23] discuss the effect of constrained interactions in their model. In their discussion, when the constraint of actively linking number is relatively smaller than the overall population, payoff dominate convention will be selected finally, and when the limit number is large enough, risk dominate convention can be observed in long run. The main difference between our model and the model in Staudigl and Weidenholzer [23] is that there are two groups with different constraints of actively linking number in our model.

There is a different branch of literature where in those model agents not only decide which action to play in the coordination game, but also choose among several locations in which they interact with other agents and play the coordination game. The previous works in this branch include Ely [14], Bhaskar and Vega-Redondo [6], or Jackson and Watts [20]. In these literature, agents can simply move away from one location to another instead of changing their actions to coordinate with interaction partners for maximizing their respected payoffs. Therefore, by "voting by feet," agents most likely select payoff dominate convention. Dickmann [9] proposed a model where there are multiple locations and barriers on mobility through different locations. In his discussion, compared with unrestricted mobility where only payoff dominate convention emerge in the long run, if there are constraints on mobility, both two different conventions are co-existent. In Anwar [4], this result is reinforced.

The present works are also related to the recent literature in the discussion of learning through imitation of successful behavior. In the early literature (see, e.g., Ellison and Fudenberg [12, 13], or Bala and Goyal [5]), in the scenario where the information received by agents is only from their own past experience and the experience of neighbors, the overall population will select efficient outcomes in the long run. Later Alós-Ferrer and Weidenholzer [1] show that payoff dominate convention will emerge uniquely in the scenario where the interactions are neither too global nor too local, based on the circular city model in Ellsion [10] and learning by imitation rules. Alós-Ferrer and Weidenholzer [3] expand their discussion to arbitrary networks in a local interaction model. As in Alós-Ferrer and Weidenholzer [1], agents learn by imitation of neighbors' successful actions. They show that if information on successful actions is capable of streaming through the network, as well as the minimal number of neighbors is relatively small compared with the maximal number of neighbors without interactions, efficient conventions will be selected ultimately. In Fosco and Mengel [15], they consider a Prisoners' Dilemma model where agents learn by imitation of successful behaviors of both action choices and interaction partners. They find that in the long run, defectors and cooperators both exist, yet two types of agents are fully separated into two disconnected components. The main difference between our model and these models in this branch of literature is in the revision process of agents' strategies, we consider a model of myopic best response rules à la Kandori et al. [22] and Young [25] instead of learning by imitation rules.

11.3   The Model

Consider N agents who play a 2×2 coordination games with each other. Each agent i can choose an action $a_i \in \{A, B\}$. We denote by $u(a_i, a_j)$ the payoff agent i receives from the interaction with agent j. The following table is the payoff matrix of this coordination game.

	A	B
A	a, a	c, d
B	d, c	b, b

Assume that $b > c$ and $a > d$, so that the strategy (A, A) and (B, B) are strict Nash-Equilibria. We assume $b > a$, so (B, B) is Pareto-Efficient in the sense that the payoffs are larger. Further assume that $a + c > b + d$, hence (A, A) is risk-dominate, according to Harsanyi and Selton [17]. In addition we assume that $a > c$, so that A-player will be more willing to interact with A-players. Thus, the payoffs in the coordination game is ordered in $b > a > c > d$.

Besides action choices, agents also decide whom to interact with. We denote $g_{ij} = 1$ if agent i actively forms a link to agent j. Otherwise $g_{ij} = 0$. Agents cannot link to themselves, so we have $g_{ii} = 0$ for all $i \in I$. We divide the entire population into two groups N_1 and N_2, and denote by $n_1 = |N_1|$ and $n_2 = |N_2|$ the number of population in two groups where $N = n_1 + n_2$. Every agent in N_1 can only support a limited number k_1 of active links and every agent in N_2 can only support at most k_2 active links, where $1 \leq k_1 < k_2 \leq N - 1$. And there is no restriction in the number of incoming links for each agent in I. One linking strategy of agent i can be summarized by an N-tuple $g_i = (g_{i1}, g_{i2}, \ldots, g_{in_1}, \ldots, g_{i(n_1+1)}, \ldots, g_{iN}) \in \mathcal{G}_i = \{0, 1\}^N$.

A pure strategy s_i of agent i includes two parts: action choice $a_i \in \{A, B\}$ and linking strategy $g_i \in \mathcal{G}_i$. Hence, $s_i = (a_i, g_i) \in \mathcal{S}_i = \{A, B\} \times \mathcal{G}_i$. We denote the strategy profile by a tuple $\mathcal{S}$, where $s = (s_i)_{i \in I} \in \mathcal{S} = \Pi_{i \in I}$. We assume that $d_i^{out} = \Sigma_j g_{ij}$ is the number of players i actively links to, and $d_i^{in} = \Sigma_j g_{ji}$ is the number of players he passively links to. Hence, $\forall i \in N_1$, we have $d_i^{out} \leq k_1$ and $\forall j \in N_2$, we have $d_j^{out} \leq k_2$. Further, we assume that m is the number of A-players in the whole group I at the given strategy profile s. Hence, the number of B-players is $N - m$.

We assume that payoff only flows to the active side of a link. And the payoff of a agent is determined by the sum of payoffs he received from the coordination games, minus the entire linking cost γ. So given the strategy profile $s = (s_i)_{i \in I}$, the overall payoff is given by:

$$U_i(s, s_{-i}) = \sum_{j=1}^{N} g_{ij} u(a_i, a_j) - \gamma d_i^{out} \tag{11.1}$$

The first term in Function 11.1 is the total payoffs he received from the coordination games with his neighbors he actively links with. The second term is the total linking costs. In a more general model, there is a third term that means the payoffs from the coordination games with agents he passively linked with. So the payoff function should be:

$$\tilde{U}_i(s, s_{-i}) = \sum_{j=1}^{N} g_{ij} u(a_i, a_j) - \gamma d_i^{out} + p(d_i^{in}) \tag{11.2}$$

In Function 11.2, $p(d_i^{in})$ is an increasing function depending on the number of passive connections d_i^{in}, but not on the actions used by those neighbors. And since the payoff of the coordination games is only on the active side, agents need not to consider the passive links when they make a decision on action choice and linking strategy. Hence, for a best responding agent, Function 11.1 equals to Function 11.2.

In the following, we denote by:

$$\overrightarrow{A_i[k_1]B_j[k_2]} = \{s \in \mathscr{S} | a_i = A, a_j = B \text{ and } d_i^{out} = k_1, d_j^{out} \quad (11.3)$$

$$= k_2, \forall i \in N_1, j \in N_2\}$$

the set of monomorphic states, where all agents in group N_1 choose the same action A and support k_1 links, moreover, all agents in group N_2 choose the same action B and support k_2 links. We consider all states in this set as absorbing sets.

We assume that time is discrete, denoted by $t = 1, 2, 3, \ldots$. At each period t, the strategy profile of all agents is determinate. But at each period t, each agent has a chance to modify their strategy with an independent probability $\lambda \in (0, 1)$. When such a modification opportunity arises, each agent chooses a best response to the other agents' strategies to maximize her payoff in the preceding period. More formally, the period t, agent i chooses the strategy:

$$s_i(t) \in arg \max_{s_i \in \mathscr{S}_i} U_i(s_i, s_{-i}(t-1)) \quad (11.4)$$

where $s_{-i}(t-1)$ indicates the strategy profile of the other agents except i in the previous period. If agents have multiple best response strategies, we assume that they choose one of them at random. We assume that each agent may make mistakes that she updates her strategy—including action or links—at random, with a positive probability ε. We further assume that ε is independent across agents, time, and payoffs.

Note that agents make a decision of action and linking choice only depending on the distribution of actions in the overall population. Since the action choices and linking decision is synchronous, an agent has to analyze which particular action she will choose taking into account her optimal linking decision. Therefore, the decision problem is divided into two steps: First, given the distribution of both actions A and B, decide the optimal linking set. Second, given the optimal set in the first step, determine which action to play. To solve the question in the first step, we use the concept of *link optimized payoff function*, for short LOP given by Staudigl and Weidenholzer [23]. The LOP is given by:

$$v(a_i, m) = \max_{g_i \in \mathscr{G}_i} \tilde{U}((a_i, g_i), m) \quad (11.5)$$

where $\tilde{U}((a_i, g_i), m)$ denotes the payoff of agent i when she chooses strategy $s_i = (a_i, g_i)$ and the total number of A-players is m. Given the LOP, we can solve the problem in the second step. Assume that agent i played action $a_{i,t-1}$ in the period $t-1$, and meanwhile there were m_{t-1} agents playing action A. The action in period t played by agent i is determined in the following way.

First, if $a_{i,t-1} = A$, switch to B when $v(B, m_{t-1} - 1) > v(A, m_{t-1})$, randomize between A and B when $v(B, m_{t-1} - 1) = v(A, m_{t-1})$, and stay with A otherwise.

Second, if $a_{i,t-1} = B$, switch to A when $v(A, m_{t-1} + 1) > v(B, m_{t-1})$, randomize between A and B when $v(A, m_{t-1} + 1) = v(B, m_{t-1})$, and stay with B other wise.

11.4   Constrained Interactions

We start our discussion by analyzing the optimal linking strategies of agents on a scenario of low linking cost ($0 \leq \gamma \leq d$). Note that because of the constrained condition $a > c$, an A-player is more willing to interact with another A-player over with a B-player. Hence, an A-player will link with other A-players first. Only in the case that the limited number of active links do not less than the total number of A-players, an A-player will consider to link up to B-players after they have already formed links to all other A-players. Similarly, the constraint $b > d$ implied that B-players will first choose to interact with other B-players. And for the reason that the linking cost is low, all agents will first link up to all agents with the same action and then if there were slots remained, agents would fill up the slots with other agents of different kind. Consequently, given a distribution of actions $(m, N - m)$, the LOPs of any A-player and B-player in N_1 are given by:

$$v_i(A, m) = a \min\{k_1, m - 1\} + c(k_1 - \min\{k_1, m - 1\}) - \gamma k_1 \qquad (11.6)$$

$$v_i(B, m) = b \min\{k_1, N - m - 1\} + d(k_1 - \min\{k_1, N - m - 1\}) - \gamma k_1 \quad (11.7)$$

Similarly, $\forall j \in N_2$, the LOPs of A-players and B-players are given by:

$$v_j(A, m) = a \min\{k_2, m - 1\} + c(k_2 - \min\{k_2, m - 1\}) - \gamma k_2 \qquad (11.8)$$

$$v_i(B, m) = b \min\{k_2, N - m - 1\} + d(k_2 - \min\{k_2, N - m - 1\}) - \gamma k_2 \quad (11.9)$$

An A-player will switch to action B with positive probability if $v(B, m - 1) \geq v(A, m)$. Depending on the relationship between m, N, k_1 and k_2, we have seven sub-cases to analyze when solving the problem of switching thresholds for A-players. Similarly, a B-player will switch to A if $v(A, m + 1) \geq v(B, m)$, and there are seven sub-cases of switching thresholds for B-players as well. We report our results of switching thresholds in Table 11.1 as well. With the help of Table 11.1 we can characterize all absorbing sets.

Table 11.1 Switching thresholds, where "a.s" means that a player always switches to the other action and "n.s" means that a player never switches to the other action

Switching thresholds for A-players

$v(B, m-1) \geq v(A, m)$		$k_2 > k_1 \geq m-1$ $\;$ $k_2 > k_1 \geq N-m$	$k_2 \geq m-1 > k_1$ $\;$ $k_2 > k_1 \geq N-m$	$m-1 > k_2 > k_1$ $\;$ $k_2 > k_1 \geq N-m$
$0 \leq \gamma \leq d$	v_i	$m \leq \dfrac{(N-1)(b-d)-k_1(c-d)}{a+b-c-d}+1$	$m \leq N-\frac{a-d}{b-d}k_1$	$m \leq N-\frac{a-d}{b-d}k_1$
	v_j	$m \leq \dfrac{(N-1)(b-d)-k_2(c-d)}{a+b-c-d}+1$	$m \leq \dfrac{(N-1)(b-d)-k_2(c-d)}{a+b-c-d}+1$	$m \leq N-\frac{a-d}{b-d}k_2$
		$k_2 > k_1 \geq m-1$ $\;$ $k_2 \geq N-m > k_1$	$k_2 \geq m-1 > k_1$ $\;$ $k_2 \geq N-m > k_1$	$m-1 > k_2 > k_1$ $\;$ $k_2 \geq N-m > k_1$
$0 \leq \gamma \leq d$	v_i	a.s.	a.s.	a.s.
	v_j	$m \leq \dfrac{(N-1)(b-d)-k_2(c-d)}{a+b-c-d}+1$	$m \leq \dfrac{(N-1)(b-d)-k_2(c-d)}{a+b-c-d}+1$	$m \leq N-\frac{a-d}{b-d}k_2$
			$N-m > k_2 > k_1$	
$0 \leq \gamma \leq d$	v_i	a.s.	a.s.	a.s.
	v_j	a.s.	a.s.	a.s.

(continued)

Table 11.1 (continued)

Switching thresholds for B-players

$v(A, m+1) \geq v(B, m)$		$k_2 > k_1 > m$	$k_2 > m \geq k_1$	$m \geq k_2 > k_1$
$0 \leq \gamma \leq d$	v_i	$k_2 > k_1 > N-m-1$ $m \geq \dfrac{(N-1)(b-d)-k_1(c-d)}{a+b-c-d}$	$k_2 > k_1 > N-m-1$ $m \geq N-1-\dfrac{a-d}{b-d}k_1$	$k_2 > k_1 > N-m-1$ $m \geq N-1-\dfrac{a-d}{b-d}k_1$
	v_j	$m \geq \dfrac{(N-1)(b-d)-k_2(c-d)}{a+b-c-d}$	$m \geq \dfrac{(N-1)(b-d)-k_2(c-d)}{a+b-c-d}$	$m \geq N-1-\dfrac{a-d}{b-d}k_2$
		$k_2 > k_1 > m$ $k_2 > N-m-1 \geq k_1$	$k_2 > m \geq k_1$ $k_2 > N-m-1 \geq k_1$	$m \geq k_2 > k_1$ $k_2 > N-m-1 \geq k_1$
$0 \leq \gamma \leq d$	v_i	n.s.	n.s.	n.s.
	v_j	$m \geq \dfrac{(N-1)(b-d)-k_2(c-d)}{a+b-c-d}$	$m \geq \dfrac{(N-1)(b-d)-k_2(c-d)}{a+b-c-d}$ $N-m-1 \geq k_2 > k_1$	$m \geq N-1-\dfrac{a-d}{b-d}k_2$
$0 \leq \gamma \leq d$	v_i	n.s.	n.s.	n.s.
	v_j	n.s.	n.s.	n.s.

Lemma 1 *The sets* $\overrightarrow{A[k_1]A[k_2]}$ *and* $\overrightarrow{B[k_1]B[k_2]}$ *and* $\overrightarrow{B[k_1]A[k_2]}$ *are the three absorbing sets.*

Proof In each case in Lemma 1, if an agent's optimal choice is to remain her action, then it is optimal for agents with different action to switch. Consider a state $s \in$ $\overrightarrow{A[k_1]A[k_2]} \cup \overrightarrow{B[k_1]B[k_2]} \cup \overrightarrow{B[k_1]A[k_2]}$, a revising agent i with action a_i will always remain that action. Further, consider that for each pair of states $s, s' \in \overrightarrow{A[k_1]A[k_2]}$ (and also for each pair in $\overrightarrow{B[k_1]B[k_2]}$ or $\overrightarrow{B[k_1]A[k_2]}$), in the best response process ties are broken randomly. There is a positive probability of moving from state s to state s' without mutation. Then the states in $\overrightarrow{A[k_1]A[k_2]}$ (and also in $\overrightarrow{B[k_1]B[k_2]}$ or $\overrightarrow{B[k_1]A[k_2]}$) form an absorbing set.

Further, consider any state $s \notin \overrightarrow{A[k_1]A[k_2]} \cup \overrightarrow{B[k_1]B[k_2]} \cup \overrightarrow{B[k_1]A[k_2]}$, for each group N_i of agents, with positive probability they will choose the same action and move to a state $\overrightarrow{A[k_i]}$ (or $\overrightarrow{B[k_i]}$) according to Staudigl and Weidenholzer [23]. Then for the entire group I, they will move to a state of $\overrightarrow{A[k_1]A[k_2]}$, $\overrightarrow{B[k_1]B[k_2]}$, $\overrightarrow{A[k_1]B[k_2]}$ and $\overrightarrow{B[k_1]A[k_2]}$ with positive probability. Consider that $n_1 < n_2$, we find that any states in $\overrightarrow{A[k_1]B[k_2]}$ will finally move to $\overrightarrow{B[k_1]B[k_2]}$ without mutation. After that, it follows that any state $s \notin \overrightarrow{A[k_1]A[k_2]} \cup \overrightarrow{B[k_1]B[k_2]} \cup \overrightarrow{B[k_1]A[k_2]}$ will reach to a state $s' \in \overrightarrow{A[k_1]A[k_2]} \cup \overrightarrow{B[k_1]B[k_2]} \cup \overrightarrow{B[k_1]A[k_2]}$ ultimately.

11.5 Conclusion

We have developed a model of social coordination and network formation with heterogeneous constraints, where agents are divided into two groups and supports different constraint numbers of active links. Compared to the conclusion in Staudigl and Weidenholzer [23], when constraints are homogeneous, the co-existence of payoff dominate and risk- dominate absorbing sets at same period in long-run will not emerge. However, in our discussion, when constraints are heterogeneous, the co-existence of both absorbing set at same period is possible. We find that in the situation of low linking costs, if the number of agents choosing efficient action is relatively larger than the constrained linking number of agents in both two groups, $\overrightarrow{B[k_1]B[k_2]}$ will be the only absorbing set. If the number of agents choosing efficient action is larger than the limited linking number of agents in group N_1 and smaller than the limited linking number of agents in group N_2, the co-existence of $\overrightarrow{B[k_1]B[k_2]}$ and $\overrightarrow{B[k_1]A[k_2]}$ is observed. Otherwise, the set $\overrightarrow{A[k_1]A[k_2]}$ is one of the absorbing sets.

We can extend our research to many natural aspects. First, it is acceptable to do a further research on the long run equilibria as well as discuss the difference between cases where constraints are homogeneous and heterogeneous. Second, as in Staudigl and Weidenholzer [23], we could consider the scenario where the cost of active links is high and study the diversification of absorbing sets.

References

1. Alós-Ferrer, C., Weidenholzer, S.: Imitation, local interactions, and efficiency. Econ. Lett. **93**, 163–168 (2006)
2. Alós-Ferrer, C., Weidenholzer, S.: Partial bandwagon effects and local interactions. Games Econ. Behav. **61**, 1–19 (2007)
3. Alós-Ferrer, C., Weidenholzer, S.: Contagion and efficiency. J. Econ. Theory **143**, 251–274 (2008)
4. Anwar, A.W.: On the co-existence of conventions. J. Econ. Theory **107**, 145–155 (2002)
5. Bala, V., Goyal, S.: A noncooperative model of network formation. Econometrica **5**(68), 1181–1229 (2000)
6. Bhaskar, V., Vega-Redondo, F.: Migration and the evolution of conventions. J. Econ. Behav. Organ. **13**, 397–418 (2004)
7. Blume, L.: The statistical mechanics of strategic interaction. Games Econ. Behav. **5**, 387–424 (1993)
8. Blume, L.: The statistical mechanics of best-response strategy revision. Games Econ. Behav. **11**, 111–145 (1995)
9. Dieckmann, T.: The evolution of conventions with mobile players. J. Econ. Behav. Organ. **38**, 93–111 (1993)
10. Ellison, G.: Learning, local interaction, and coordination. Econometrica **61**, 1047–1071 (1993)
11. Ellison, G.: Basins of attraction, long-run stochastic stability, and the speed of step-by-step evolution. Rev. Econ. Stud. **67**, 17–45 (2000)
12. Ellison, G., Fudenberg, D.: Rules of thumb for social learning. Eur. J. Polit. Econ. **101**, 612–643 (1993)
13. Ellison, G., Fudenberg, D.: Word-of mouth communication and social learning. Q. J. Econ. 110:95–126 (1995)
14. Ely, J.C.: Local conventions. Adv. Theor. Econ. **2**, 1–30 (2002)
15. Fosco, C., Mengel, F.: Cooperation through imitation and exclusion in networks. J. Econ. Dyn. Control. **35**, 641–658 (2011)
16. Goyal, S., Vega-Redondo, F.: Network formation and social coordination. Games Econ. Behav. **50**, 178–207 (2005)
17. Harsanyi, J., Selten, R.: A General Theory of Equilibrium Selection in Games. The MIT Press, Cambridge (1988)
18. Hojman, D., Szeidl, A.: Endogenous networks, social games, and evolution. Games Econ. Behav. **55**, 112–130 (2006)
19. Jackson, M.O., Watts, A.: On the formation of interaction networks in social coordination games. Games Econ. Behav. **41**, 265–291 (2002)
20. Jackson, M.O., Watts, A.: Social games: matching and the play of finitely repeated games. Games Econ. Behav. **70**, 170–191 (2010)
21. Kandori, M., Rob, R.: Evolution of equilibria in the long run: a general theory and applications. J. Econ. Theory **65**, 383–414 (1995)
22. Kandori, M., Mailath, G.J., Rob, R.: Learning, mutation, and long run equilibria in games. Econometrica **61**, 29–56 (1993)
23. Staudigl, M., Weidenholzer, S.: Constrained interactions and social coordination. J. Econ. Theory **152**, 41–63 (2014)
24. Weidenholzer, S.: Coordination games and local interactions: a survey of the game theoretic literature. Games **1**(4), 551–585 (2010)
25. Young, P.: The evolution of conventions. Econometrica **61**, 57–84 (1993)

Chapter 12
Stable and Efficient Structures for the Content Production and Consumption in Information Communities

Larry Yueli Zhang and Peter Marbach

12.1 Introduction

Communities are an important structure that widely exists in real-world online and offline social networks. A common type of community is the *information community* in which the members of the community produce content and consume the content produced by other members, with the most popular example being Reddit [10] where each "subreddit" is essentially an information community with a specific topic of interest. Real-world communities often exhibit inherent structures such as the high density of interactions within the community and the existence of a core set of active members who would contribute the majority of the content in the community ("the Law of the Few") [4, 12]. For example, in [12] it is empirically shown that, in an online forum, only 12% of the users are actively generating content while the majority of the users are almost silent. There have been a large body of research work on community detection algorithms based on such structures. However, there is still a lack of the formal understanding of why these structures would consistently and naturally emerge during the formation process of real-world communities. Understanding the formation process of these natural structures is important as it provides us a microscopic view of the working mechanisms of communities and would enable us to utilize communities more efficiently.

Our overall hypothesis is the following: real-world social network structures have been going through an evolutionary process, and as a result of that only the optimal structure (in terms of stability and efficiency) can survive, sustain therefore exist widely in real-world social networks. In other words, if we observe a widely

L. Y. Zhang (✉) · P. Marbach
University of Toronto, Toronto, ON, Canada
e-mail: ylzhang@cs.toronto.edu; marbach@cs.toronto.edu

© Springer Nature Switzerland AG 2019
J. B. Song et al. (eds.), *Game Theory for Networking Applications*,
EAI/Springer Innovations in Communication and Computing,
https://doi.org/10.1007/978-3-319-93058-9_12

existing structure in real-world social networks, then this structure must be optimal in the sense that it has stable user behaviours and it is efficient for the purpose of the network.

In the case of information communities, each member in the community is an agent who can choose to spend certain portions of their time in producing content items or in consuming content produced by other members. In order for the community to be stable, all members' time allocation strategies should collectively form a Nash equilibrium, i.e., each member would get penalized by deviating from the equilibrium strategy. A member in the community can be rewarded by either production or consumption. For consumption, the rewarded is from the consumed content itself; for production, a member is rewarded when the content she produces is consumed by other members of the community (the reputation effect). A community structure is called "efficient" when it can provide its members the highest possible amount of reward. If we use a mathematical model to formulate the above behaviours and efficiency measures, we will then be able to formally analyse the condition under which the community structure is optimally stable and efficient, therefore obtain a mathematical description of the "surviving and sustaining" community structure. The validity of the model would be verified if the result of the analysis happens to agree with the widely existing structures observed in real-world communities. Compared to the empirical observations, the formal analytical results would provide us more refined understanding of the microscopic working mechanisms of the real-world communities.

In this paper, we formulate a model that captures the production and consumption behaviours inside an information community. Our analysis results show that the structure with a small set of "celebrity producers" is the optimally stable and efficient structure. These analysis results provide possible explanations to the sociological observations such as "the Law of the Few" and also provide insights into how to effectively build and maintain the structure of information communities.

12.2   Related Works

Social network analysis has been one of the fastest growing research fields in the twenty-first century. We refer readers to Scott et al. [11] for a comprehensive coverage of the development of the subject, rather than listing the large collection of references in this paper. Experimental works observed interesting properties of real-world complex networks such as the power-law degree distribution, the small-world phenomena and the community structure. These observations lead to modelling works that tried to explain why the observed properties would emerge, such models include the preferential attachment models, the copying model and the forest fire model. However, most of these works were studying macroscopic structural properties rather than looking into the internal microscopic structures of the network.

The community structure has been an interesting topic for researcher in the field of social network analysis. A large body of work has been devoted to modelling and detecting community structures in large scale social networks (e.g., [2, 5, 6, 8, 9]). The networks are often represented by graphs in which the vertices represent underlying social entities and the edges represent some sort of social tie or interaction between pairs of vertices. Our model differs in the sense that it also considers the user behaviours on top of the network connections.

In [1], the efficiency of a network in terms of information diffusion is studied, a mathematical analysis is performed to investigate the optimal network structure to achieve the best efficiency for information diffusion (high precision, high recall and low diameter), and the result shows that a Kronecker-graph [6] would satisfy such conditions. The approach taken in [1] is similar to the approach we take in this paper except that we are more focussed on the community related aspects. The work in [3] used a game-theoretic model to study the emergence of the "Law of the Few" but it is also in the context of information diffusion rather than about communities. The work in [7] is the closest to the interest of this paper. In [7], a game theoretic model is formulated to analyse the community structures in terms of content production and consumption. Each member's strategy involves choosing a particular interest to produce or consume content on. The result shows that in the Nash equilibrium of the model the members' choices form community structures. The difference of our work from [7] is that we focus on the internal structure of a single community rather than on the scale of multiple communities, and besides the Nash equilibrium, we also take the social welfare into consideration.

12.3 Model

We will first describe the general configuration of the model and the payoff/reward functions, then in Sects. 12.3.2 and 12.3.3, we introduce two variations of modelling the internal relations between the community members. Both models will be analysed and the results will be compared in Sect. 12.4.

12.3.1 General Configuration

We have a single community with n members indexed by $1 \leq i \leq n$. Each member is capable of both producing and consuming content items. The produced content items could be chosen by all members or a subset of the members of the community for consumption. Each member has a limited total amount of time which could be allocated to either production or consumption, and each member makes a decision about how much of their time to allocate to production and consumption. A member is rewarded if their products are consumed by members of the community (the production reward), or if the member consumes an item that is produced by a member of the community (the consumption reward). Each member's objective is to maximize their total individual reward from both production and consumption.

The time slot: In our model, we investigate everything that happens within a *unit time*. The assumption is that the long-term behaviour of a member is the repetition of their behaviour within a unit time.

Rates of content production and consumption: We define $N_p \geq 0$ to be the number of content items that a member can produce if they were to spend 100% of their unit time on production; and we let $N_c \geq 0$ be the number of items that a member can consume within a unit time slot if they were to spend 100% of their time on consumption. We assume that all members share the same values of N_p and N_c, and we assume the following inequality:

$$0 \leq \frac{N_c}{n N_p} \leq 1 \tag{12.1}$$

This assumption is reasonable because $n N_p$ is the largest possible number of content items that can be produced, an N_c value that is larger than $n N_p$ would be unrealistic.

A member's time allocation strategy: Let α_i ($0 \leq \alpha \leq 1$) be the portion of the unit time that member i allocates to production (therefore $1 - \alpha$ is allocated to consumption). Each member chooses their own α_i, we will investigate if a set of choices of α_i would lead to a Nash equilibrium. Within the unit time, a member can consume at most $(1 - \alpha_i) \cdot N_c$ items. If the number of available items is less than or equal to this number, then each member would consume all the available items without any choice; if the total number of available items is greater than this number, then the member would choose a subset (of size $(1 - \alpha_i) \cdot N_c$) of the available items to consume, uniformly at random.

The *production reward* models the "reputation effect" in social networks, i.e., having content products consumed by other people is rewarding for the producer of the content. The reward for each item that a member produces is proportional to the number of members who consume the item, with a constant factor r_p, i.e., if an item is consumed by m members, then the reward for this item is $r_p \cdot m$. The total production reward for a member is the sum of the rewards of all items that the member produces. The constant factor r_p is the same for all members. The *consumption reward* of a given item is a constant r_c. The total consumption reward of a member is r_c multiplied by the number of items consumed by the member. The total individual reward of a member in the community is the sum of their production reward and consumption reward. The sum of the total individual rewards of all members in the community is the *social welfare*. While each member tries to maximize their own individual reward, the overall efficiency of the community is measured by its social welfare.

The following two subsections will define two variations of the internal relational structure of the community.

12.3.2   The Celebrity-Follower Community Structure

Under the celebrity-follower relational structure, a subset of the members of the community are "celebrities" that are followed by everyone in the community, i.e., the content items produced by a celebrity member can be seen by all members of the community. A non-celebrity member has zero followers, i.e., an item produced by a non-celebrity member cannot be seen or consumed by any member.

Let η be the portion of celebrity members, i.e., the number of celebrity members is ηn. When $\eta = 1$, all members are connected via a complete graph. When η is small, we have a small core of celebrities that would be responsible for producing all content items in the community. If visualized as a directed graph, the structure would have $\eta \cdot n^2$ edges in total. Note that we are assuming a member can be a follower of themselves so the graph can have self-pointing edges. This would lead to cleaner analysis results.

Note that we are not making any assumptions about how large the value of η is, and it is interesting to see whether the efficiency of the community system can be different with η's value being in different ranges. In real-world communities, we often observe patterns that are similar to the celebrity-follower structure, i.e., a small subset of "elite contributors" would produce most of the content items that are consumed by all members of the community, and a community typically has a significant portion of "lurkers." We will be able to provide a theoretical explanation to this real-world phenomenon.

12.3.3   The Uniform Community Structure

In contrast to the celebrity-follower structure where the members play unequal roles in the community, the uniform relational structure has all members with the equal role, i.e., every member has the same number of followers and follows the same number of other members. In terms of a graph, it is a regular graph where every vertex has the same in-degrees and out-degrees.

To make this structure comparable with the celebrity-follower structure, we let it have the same number of edges as the celebrity-follower graph. The celebrity-follower graph discussed in the previous section has $\eta \cdot n^2$ edges, therefore, in the uniform graph, we let each vertex have in-degree $\eta \cdot n$ as well as out-degree $\eta \cdot n$.

12.3.4   Summary of the Model

Overall, our model is a game-theoretic model where each agent (member of the community) chooses a strategy (α_i) with the objective of optimizing their individual reward. The efficiency of the whole community is measured by the social welfare (total reward of all members). The stability of the community is indicated by whether the strategies of all members collectively form a Nash equilibrium.

12.4 Analysis

Our hypothesis is that, in order to exist and sustain in the real world, a social structure must be stable and efficient. For an information community, this means that the members' strategies form a Nash equilibrium while the social welfare of the community is maximized. Therefore, our analysis will take the following approach: we first derive the set of members' strategies that would maximize the social welfare of the community, then we investigate the condition for this set of strategies to form a Nash equilibrium.

In Sects. 12.4.1 and 12.4.2, we perform the analyses for the celebrity-follower and uniform structures, respectively, then we will compare and discuss the analysis results.

12.4.1 Analysis of the Celebrity-Follower Structure

The following theorem summarizes the analysis results for communities with the celebrity-follower structure.

Theorem 1 *For a community with the celebrity-follower structure where there are $\eta \cdot n$ celebrity members, the maximum social welfare and the Nash equilibrium are described in the following cases.*

Case 1 If $\eta < min(\frac{N_c}{nN_p}, 1 - \frac{N_c}{nN_p}, 1 - \frac{N_c r_c}{nN_p r_p})$, then the maximum social welfare is reached when a member i of the community takes the following strategy:

$$\alpha_i = \begin{cases} 1 & \text{if member i is a celebrity} \\ 0 & \text{otherwise} \end{cases} \tag{12.2}$$

The maximum social welfare G_{max} is the following:

$$G_{max} = \eta(1 - \eta)n^2 N_p(r_p + r_c) \tag{12.3}$$

*This set of strategies **always** forms a Nash equilibrium under this case.*

Case 2 If $\frac{1}{2} < \frac{N_c}{nN_p} \leq 1$ and $1 - \frac{N_c}{nN_p} \leq \eta \leq \frac{N_c}{nN_p}$, then the maximum social welfare is reached when a member i of the community follows the following strategy.

$$\alpha_i = \begin{cases} \frac{N_c}{N_c + \eta n N_p} & \text{if member } i \text{ is a celebrity} \\ 0 & \text{otherwise} \end{cases} \qquad (12.4)$$

The maximum social welfare G_{max} is the following:

$$G_{max} = \frac{\eta n^2 N_c N_p (r_p + r_c)}{N_c + \eta n N_p} \qquad (12.5)$$

*However, this set of strategies **never** forms a Nash equilibrium under this case.*

Case 3 If $0 \le \frac{N_c}{n N_p} \le \frac{1}{2}$ and $\frac{N_c}{n N_p} \le \eta \le 1 - \frac{N_c}{n N_p}$, then the maximum social welfare is reached when a member i of the community follows the following strategy.

$$\alpha_i = \begin{cases} \frac{N_c}{\eta n N_p} & \text{if member } i \text{ is a celebrity} \\ 0 & \text{otherwise} \end{cases} \qquad (12.6)$$

The maximum social welfare under this strategy is

$$G_{max} = N_c \left(n - \frac{N_c}{N_p} \right) (r_p + r_c) \qquad (12.7)$$

*This set of strategies **never** forms a Nash equilibrium under this case.*

Case 4 $\eta > max(\frac{N_c}{n N_p}, 1 - \frac{N_c}{n N_p})$, the social welfare is maximized when a member i of the community follows the following strategy.

$$\alpha_i = \begin{cases} \frac{N_c}{N_c + \eta n N_p} & \text{if member } i \text{ is a celebrity} \\ 0 & \text{otherwise} \end{cases} \qquad (12.8)$$

The maximum social welfare under this strategy is

$$G_{max} = \frac{\eta n^2 N_c N_p (r_p + r_c)}{N_c + \eta n N_p} \qquad (12.9)$$

*This set of strategies **never** forms a Nash equilibrium under this case.*

The detailed proof of Theorem 1 can be found in the appendix of [13]. This theorem shows that Case 1 is the only case where the members' strategies reach a Nash equilibrium while the social welfare is maximized. In other words, in order for the community to be optimally stable and efficient, the portion of celebrity members must be small enough, i.e., $\eta < min(\frac{N_c}{n N_p}, 1 - \frac{N_c}{n N_p}, 1 - \frac{N_c r_c}{n N_p r_p})$.

12.4.2 Analysis of the Uniform Structure

The following theorem summarizes the analysis results for communities with the celebrity-follower structure.

Theorem 2 *For a community with the uniform structure where each member has ηn followers and follows ηn members, the maximum social welfare is reached when the following set of strategies is applied.*

$$\alpha_i = \frac{N_c}{N_c + \eta n N_p} \quad \forall 1 \leq i \leq n \tag{12.10}$$

The maximum social welfare G_{max} is the following:

$$G_{max} = \frac{\eta n^2 N_p N_c (r_c + r_p)}{N_c + \eta n N_p} \tag{12.11}$$

The above set of strategies forms a Nash equilibrium if and only if the following condition is true.

$$\eta \leq \left(\frac{N_c r_c}{n N_p r_p} + \frac{1}{n} \right) \tag{12.12}$$

The detailed proof of Theorem 2 can be found in the appendix of [13]. This result shows that, assuming the uniform community structure, there exist a simple set of strategies that is stable while the social welfare is maximized. What we are interested in is how the optimal efficiency of the uniform structure compares with that of a community with the celebrity-follower structure. The following theorem provides us a formal result.

Theorem 3 *Let $G_{max\text{-}celebrity}$ be the maximum social welfare with a Nash equilibrium for the celebrity-follower community structure (Eq. (12.3)) and $G_{max\text{-}uniform}$ be the maximum social welfare with a Nash equilibrium for the uniform community structure (Eq. (12.11)). The following is always true:*

$$G_{max\text{-}celebrity} \geq G_{max\text{-}uniform} \tag{12.13}$$

The detailed proof of Theorem 3 can be found in the appendix of [13]. This theorem provides a simple and clear result: given being in its optimally stable and efficient state, a community with the celebrity-follower structure always has a better optimal social welfare than a community with the uniform structure.

12.5 Discussions

The combination of the analysis results in Sects. 12.4.1 and 12.4.2 provides us two different angles of explaining the common "law-of-the-few" structural patterns that widely exist in real-life information communities. A given community structure, in order to exist and sustain, must be both stable and efficient, meaning that the community can stably stay at the state with the maximum social welfare. Theorem 1 tells us that the community can only be stable and efficient if there is a small enough "core" of celebrity members who will actively contribute all the content to be consumed by all members of the community, while the majority of the community members would simply consume the content produced by the core members. A community structure that does not satisfy this condition would not be stable therefore would not commonly exist in reality.

Moreover, among the different possible structure that are both stable and efficient, some structures are more efficient than others. Theorem 3 shows that the small-core celebrity-follower structure is not only stable and efficient, but also it is more efficient than other stable structures such as the uniform structure.

With the above two factors taken into account, the celebrity-follower structure with a small set of celebrities becomes the winner, therefore becomes the commonly existing structure in real-world information communities.

In the equilibrium state, the strategies of the celebrity and non-celebrity members are clearly differentiated: the celebrity members should dedicate all of their time in production whereas the non-celebrity members should spend all of their time on consumption. These specialized producing and consuming behaviours also coincide with real-world observations: in a web service such as Reddit, the visitors of a typical subreddit would often separate into two different roles, i.e., the "active contributors" who frequently post content in the subreddit and the "lurkers" who would always just consume content silently.

The analysis results also provide insights into how to effectively build and maintain information communities. The most important takeaway from our analysis results is that there should be mechanisms that encourage the formation of a small-core celebrity-follower structure inside the community. For example, many online social network applications use features such as "thumb-up" or "upvote" to promote and reward high quality content that are liked by many community members. Besides providing effective content filtering (ranking by votes), this voting mechanism also encourages the optimally stable and efficient community structure: since the production reward is only earned when a post is upvoted, the members who would produce low-quality content would not be rewarded and would essentially become the non-celebrity members in the celebrity-follower structure. The members who produce high-quality content would be rewarded by the upvotes and becomes the celebrities in the community. The size of the core of celebrities will tend to be small if the display of the content in the community is ranked by popularity: most members will only consume a small portion of the top-ranked content items therefore only a small set of high quality producers would actually be rewarded and become the real core of the community. This analysis would lead to an interesting and counter-intuitive hypothesis: if the content display of the community is such that different members would see a diverse range of different items, then this would cause the formation of a larger-sized celebrity core or a uniform-like structure in the community which would make the community structure less stable. It would be interesting to empirically verify if this hypothesis is true in practice.

Another interesting insight is that, in the optimal community structure, the number of celebrity members in the core, i.e., $\eta \cdot n$, must satisfy that $\eta \cdot n < N_c/N_p$. This means that the size of the celebrities core does *not* increase as the size of the community n increases. This could be a possible reason of why we have communities in the first place: having a large number of people communicating in a single giant community is inefficient in terms of the total amount of production because it only allows a small number of core members to contribute in content production. Larger total production rate can be achieved by dividing people into different smaller communities each of which has its own core members, since the total number of people who will contribute in content production would be multiplied by the number of communities.

12.6   Conclusions

This paper attempts to obtain a formal understanding of the natural structural patterns of real-world information communities. We formulate a mathematical model that describes the generic content production and consumption behaviours in a community. The analysis result shows that the small-core celebrity-follower structure is the optimal structure that would lead to the optimally efficient and stable community. These analytical results agree with the sociological observations on real-world information communities. Besides providing a refined microscopic view of the working mechanisms of information communities, the analysis results also provide useful insights into how to better build and maintain the structure of information communities. Designing efficient mechanisms that encourage the formation of stable and efficient communities would be an interesting topic for future works.

References

1. Bosagh Zadeh, R., et al.: On the precision of social and information networks. In: Proceedings of the First ACM Conference on Online Social Networks. ACM, New York (2013)
2. Fortunato, S., Hric, D.: Community detection in networks: a user guide. Phys. Rep. **659**, 1–44 (2016)
3. Galeotti, A., Goyal, S.: The law of the few. Am. Econ. Rev. **100**(4), 1468–1492 (2010)
4. Gladwell, M.: The Tipping Point: How Little Things Can Make a Big Difference. Little, Brown, Boston (2006)

5. Kumar, R., Novak, J., Tomkins, A.: Structure and Evolution of Online Social Networks. Link Mining: Models, Algorithms, and Applications, pp. 337–357. Springer, New York (2010)
6. Leskovec, J., et al.: Kronecker graphs: an approach to modeling networks. J. Mach. Learn. Res. **11**, 985–1042 (2010)
7. Marbach, P.: The structure of communities in information networks. In: Information Theory and Applications Workshop (ITA) (2016)
8. Massoulié, L.: Community detection thresholds and the weak Ramanujan property. In: Proceedings of the Forty-Sixth Annual ACM Symposium on Theory of Computing. ACM, New York (2014)
9. Newman, M.E.J.: Community detection in networks: modularity optimization and maximum likelihood are equivalent. arXiv:1606.02319 (2016)
10. Reddit. http://www.reddit.com (2017)
11. Scott, J.: Social Network Analysis. Sage, Thousand Oaks (2017)
12. Zhang, J., Ackerman, M.S., Adamic, L.: Expertise networks in online communities: structure and algorithms. In: Proceedings of the 16th International Conference on World Wide Web. ACM, New York (2007)
13. Zhang, L.Y., Marbach, P.: Stable and efficient structures for the content production and consumption in information communities. arXiv:1801.04642 (2018)

Chapter 13
One-Player Game Based Influential Maximization Scheme for Social Cloud Service Networks

Sungwook Kim

13.1 Development Motivation

With the advance of the Internet of Things (IoT), Social Network Service (SNS) has attracted billions of Internet users from all over the world in the past few years. SNS connects people to provide online communication and collaboration environment beyond the geographic limitations. It is considered to be a representative of the new generation Internet applications, and many specialized SNSs have emerged [5].

Usually, the main goal of SNS is to seek reciprocal value creation to increase the productivity, quality, and opportunities of online services. To satisfy this goal, SNS users will expect more application services to fulfill their needs beyond fundamental service functions. Therefore, how to link the needs of users and shape the designs for better service utilization is an important issue in SNS research fields [6–16].

At present, Cloud Computing (CC) has been developed rapidly and becomes very common. Usually, cloud is used in science to describe a large agglomeration of objects that visually appear from a distance. In the Information and Communications Technology (ICT) field, it is a kind of Internet-based computing paradigm that provides shared processing resources and data to computers and other devices on demand. In particular, this paradigm represents a distributing computing model for enabling ubiquitous, convenient, on-demand network access to a shared pool of configurable computing resources. Therefore, CC technology provides flexible and scalable services without having the computing resources installed directly on SNS users' systems [15–17].

For the interoperability of the SNS and CC services, a new concept, Social Cloud (SC) was introduced based on the notion of resource and service collaboration. SC

S. Kim (✉)
Department of Computer Science, Sogang University, Seoul, South Korea
e-mail: swkim01@sogang.ac.kr

© Springer Nature Switzerland AG 2019
J. B. Song et al. (eds.), *Game Theory for Networking Applications*,
EAI/Springer Innovations in Communication and Computing,
https://doi.org/10.1007/978-3-319-93058-9_13

is a novel scalable computing model where resources are beneficially shared among a group of Social Network (SN) users. From [2], we rehearsal the formal definition of SC as—*A social cloud is a resource and service sharing framework utilizing relationships established between members of a social network.* It is a resource and service sharing framework utilizing relationships established between SNS users and CC providers. Under a dynamic IoT environment, the idea of SC has been gaining importance because of their potential for the system efficiency [10].

Despite the rapid development of the SC framework, there are some existing problems. One of the most famous problems is to maximize the social welfare. It can be likened to the influence maximization problem [3]. Consider the following scenario as an example. A CC system operator wants to provide cloud services for users of SNs. However, the CC system has a limited resource such that it can only select a small number of users to provide CC services. The CC provider wishes that these selected users would have strong relationship with their friends on the SN and share the profit of provided CC services. Therefore, through the ripple effect, a large population in the SN would be satisfied while maximizing their payoff. It's a good example of influence maximization. Simply, the influence maximization problem is the problem of detecting a set of influential users, who strongly influence the largest number of people in an SN [3].

Under widely dynamic SC system conditions, finding the best solution of the influence maximization problem is very challenging; it is an NP-hard problem [8]. In this study, we focus on the game theory and reinforcement learning algorithms to obtain an efficient solution for the influence maximization problem. Game theory is the study of strategic interactions between multiple rational game players while consistently pursuing their own objectives, which is measured in some utility scale. The importance of game theory is evident in the fact that it is now widely applied in various fields, such as economics, biology, political science, social psychology, sociology, and anthropology. Since the early 2010s, SNS and CC management issues have been added to this list [1–11]. Of course, as the critics of game theory argue, the assumption of rationality in game theory does not always hold in realistic environments. Therefore, the relaxation of the classical assumptions of game theory and the incorporation of stochasticity into game players' introspection process can provide the leverage to overcome the traditional game theory. Due to this reason, a part of game theory deals with reinforcement learning in games. Reinforcement learning in games involves modeling the processes by which players change the strategies they are using to play a game over time [1].

Motivated by the above discussion, S. Kim proposes a new SC management scheme based on the combined methodology of game theory and reinforcement learning. To effectively solve the influence maximization problem in SNs, the CC provider adaptively selects the most influential users and dynamically allocates its resource. By considering the real-world SC environments, the proposed approach needs no complete knowledge of the topological social structure. According to the combination of continuous probability distributions, the CC provider estimates each user's influential power, and adaptively selects a set of users to maximize the social welfare. This procedure imitating the interactive sequential game process is practical and suitable for real-world SC implementation. Based on the key principles of the repeated game model and reinforcement learning algorithm, the proposed scheme is implemented in realistic point of view while ensuring the system practicality.

The major contributions of the proposed scheme are: (1) the adaptive dynamics considering the current SC system environments, (2) the interactive game approach to provide an appropriate tradeoff between optimality and practicality, (3) the ability to solve effectively the real-world influence maximization problem, (4) the reciprocal combination of SNS and CC technologies, and (5) the ability to capture the reality of SC services. In this study, we pay serious attention to the practical implementation of this problem with reasonable time complexity. It is an important novelty of the proposed approach.

13.2   Related Work

Over the years, a lot of state-of-the-art research work on the influence maximization problem has been conducted. The Cloud and SNS Supported Collaboration (CSSC) scheme [7] presented novel cloud and SNS control algorithms based collaboration platform, and designed collaboration functions. These functions were ensured by privacy and permission policy and a resource allocation framework. The CSSC scheme also proposed a three-pass mechanism which can dynamically allocate virtual machines to physical machines within a low time complexity. Through the commercial implementation, the CSSC scheme can be used in real-life application [7].

The *Credit Distribution and Influence Maximization* (CDIM) scheme [4] extended the credit distribution model while incorporating the time-critical aspect of influence in SNs. In particular, this scheme described node features from different aspects and combined those components into user static influence for evaluating the original node influence. First, the CDIM scheme adopted the user dynamic influence to improve the credit assignment among adjacent nodes. And then, action propagation paths were tracked and credits were assigned after learning from the action-log and relational network structure. Finally, the CDIM scheme calculated the average marginal gain for each user and identified the users who had maximum marginal gain [4].

The Influence Maximization for Unknown Graphs (IMUG) scheme [14] proposed a heuristic algorithm for the influence maximization problem. Unlike the original influence maximization problem, this scheme assumed that the entire topological structure of the SN was not given, and only limited knowledge of the topological structure was obtained through probing. The basic idea of the IMUG scheme was greedily probing and selecting the user with the highest expected degree. As a probing strategy, snowball sampling strategy was adopted, and users were selected based on their expected degrees. Therefore, this approach cannot obtain complete knowledge of the entire topological structure of the graph. Even when knowledge of the SN topology was severely limited, the IMUG scheme can achieve a reasonable influence spread [14]. All the earlier work has attracted a lot of attention and introduced unique challenges.

13.3   Social Cloud Based Influential Maximization Algorithm

In this section, the main issue of influence maximization problem is presented. And then, for the SC system, a novel CC resource allocation algorithm is developed to maximize the social welfare. The proposed algorithm employs a reinforcement learning mechanism while considering current SC system conditions. Finally, the one-player game based algorithm in the seven-step procedures is explained in detail.

13.3.1   The Influential Maximization Problem for Social Cloud Systems

Network diffusion formulates a scenario in which local interaction along edges in a graph can generate global cascades in network state. Such diffusion processes have attracted a significant amount of recent attention. Based on the network diffusion model, a conventional influence maximization problem aims to select some nodes so that the expected number of remaining nodes influenced by selected nodes will be maximized [8–14]. In this study, we consider an SN as a directed graph $\mathbb{G} = \{\mathbf{V}, \mathbf{E}\}$ where $\mathbf{V}$ and $\mathbf{E}$ are the set of nodes and edges, respectively. Nodes represent the individual users in the SN and edges model the relationship between individual users. Mathematically, the influence maximization problem addresses the top K node set $\mathbf{S}$ from a graph $\mathbb{G}$ that satisfies

$$\mathbf{S} = \arg\max_{\mathbf{S} \subseteq \mathbf{V}}\{\mathscr{F}(\mathbf{S}, \mathbb{G})\}, s.t., |\mathbf{S}| = K \qquad (13.1)$$

where $\{\mathscr{F}(\mathbf{S}, \mathbb{G})\} : \mathbf{S} \times \mathbf{E} \to \mathbb{R}^+$ is a function of a node set $\mathbf{S}$ that provides the expected number of the influenced nodes in the graph $\mathbb{G}$. Therefore, the influence maximization problem becomes an instance of a combinatorial optimization problem. According to the underlying graph structure, $\mathbf{S}$ can be varied dynamically [11].

Nowadays, successful social media platforms are attractive not only for communication but also for information dissemination. This information diffusion in SN is regarded as an important mechanism that can improve social welfare. For example, an individual SN user needs to execute a computation complex application. Due to the limitation of embedded resource, some computation tasks can be offloaded to the CC system. When this user receives an outcome of CC service, it can be shared with his social friends. Therefore, from the viewpoint of CC provider, detecting influential users is an important issue for effective and efficient SC operations. However, the CC provider cannot perfectly know the SN's topological structure and social relationships. Therefore, the influence maximization problem is a very difficult problem [3, 7–17].

In this study, a novel SC resource allocation scheme is developed to maximize the social welfare in SN. In the proposed scheme, the cloud resource is effectively allocated while reducing the overhead of implementation complexity. To characterize the proposed scheme, there is a tuple $(\mathbb{G}, CC_p, f_{\mathbf{V}}, \mathbb{R}_{S_i}, \psi(\cdot), \mathbb{T}, \mathbb{A}, \mathcal{W}, \mathcal{T}_{S \in T}, \mathbb{U}^{CC_p})$.

1. $\mathbb{G}$ is an SN as a directed graph $\mathbb{G} = (\mathbf{V}, \mathbf{E})$. $\mathbf{V} = \{S_1 \ldots S_n\}$ is a set of SN users, and $\mathbf{E} = \{e_1, \ldots, e_m\}$ is a set of user's social relationships.
2. CC_p is a CC provider in the SC system.
3. $f_{\mathbf{V}}$ is the probability density function for the social relationship of individual user $S_i \in \mathbf{V}$; the social relationship can be measured as number of friends.
4. $\mathbb{R}_{S_i}$ is the set of user S_i's friends; If the S_j is a member of $\mathbb{R}_{S_i}$, the S_i and the S_j are connected in the SN. Based on their intimacy, $\psi(S_i, S_j)$ is denoted as the closeness of them; $\psi(S_i, S_j) \to [0, 1]$ represents the weighted connectivity value between S_i and S_j. In the proposed model, the profit sharing of CC service can be measured by $f_{\mathbf{V}}$ and ψ values. If there is no SN connection between the user S_i and S_j, $\psi(S_i, S_j) = 0$.
5. According to $\psi(\cdot)$ values, the primary influence power of S_i and $\Lambda_{S_i} = \sum_{S_j \in \mathbb{R}_{S_i}} (\psi(S_i, S_j))$.
6. $\mathbb{T}$ is a team of individual SN users selected by the CC_p to allocate the CC resource.
7. $\mathbb{A} = \{\ldots \mathscr{A}_{S_K} \ldots\}$ is a finite resource allocation set for users on being a team $\mathbb{T}$, and $\mathscr{A}_{S_K}$ means the amount of allocated CC resource for the selected $S_K \in \mathbb{T}$.
8. $\mathcal{W}$ is the CC_p's total computation resource amount for the CC services.
9. $\mathcal{T}_{S_i \in \mathbb{T}}(\mathscr{A}_{S_i})$ is the utility function to represent the social welfare generated by the S_i where $\mathcal{T}_{S_i \in T}(\mathscr{A}_{S_i}) \to \mathbb{R}^+$.
10. $\mathbb{U}^{CC_p}$ is the utility function of CC_p. It represents the total social welfare of the SC system, which is defined as $\sum_{S_k \in T} \mathcal{T}_{S_k}$.

13.3.2 Reinforcement Learning-Based Team Formation Process

When requesting the CC service, SN users ($S_{1 \leq i \leq n}$) have their social information regarding the relational connectivity. In this study, we shall assume that S_i reports individually his local connection coefficient (θ_{S_i}), which is the number of social connections. However, there is inherent uncertainty about the closeness of each connection. Naturally, the CC_p does not know the weight of each relationship; it is the private information of each S. Therefore, the CC_p's problem is to infer what are the true influence powers (Λ) of individual users from the announced information. To maximize the social welfare in the SN, the CC_p has to choose SN users to form a team ($\mathbb{T}$) whose members have higher influence powers for CC services. Therefore, the Λ-related uncertainty in the team formation problem can provide a rich agenda of challenges and questions.

The degree of the successful social welfare maximization, i.e., the outcome of the team action, would depend on the capabilities of the formed team. The choice of team members can be defined within a stochastic game model; this game-theoretic method enables the CC_p to better align his choices while maximizing the expected $\mathbb{U}^{CC_p}$. According to his own evolving knowledge, the CC_p implicitly takes into account all eventualities concerning possible team formations. More specifically, as a single game player, the CC_p maintains and updates the individual Λ values of SN users to make rational decisions regarding potential outcomes on behalf of formed teams.

When scenarios of repeated team formation activities come into consideration, the possibility of employing learning mechanisms in order to enhance the decision making of the CC_p presents itself. It is quite natural that the CC_p should be capable of exploiting the experience he gathered in the past in order to make more informed decisions. Nowadays, reinforcement learning techniques can prove to be of value to operate and interact under uncertainty. While engaging in the reinforcement learning mechanism, we can answer the question of how to make decisions that are sequentially rational [1].

In this study, S. Kim develops a one-player game model for the CC_p to take sequentially rational decisions to form teams sequentially. In particular, the proposed game model deals with learning in games. During the repeated team formation process, he suggests opportunities for the CC_p to learn about each SN user's abilities through repeated interaction, refining how teams are formed over time. Therefore, learning in games involves modeling the processes by a player, i.e., the CC_p, changes the strategies, i.e., selection of team members, over time to maximize his payoff, i.e., $\mathbb{U}^{CC_p}$.

Under the SN's uncertainty, the proposed game model effectively integrates decision making during repeated team formation. By the observation of the effects of team actions, the role of $\mathbf{CC_p}$ is to select adaptively team members who will effectively act through CC services. This situation leads us to develop an expectation mechanism to estimate the influence power (Λ) of each SN user. Based on the principle from reinforcement learning algorithms, the proposed approach relaxes the assumption that all information is completely known to predict the outcome of the uncertainty. In the proposed scheme, the $\mathbf{S_i}$'s influence power ($\Lambda_{\mathbf{S_i}}$), it is called the $\mathbf{S_i}$'s type, is defined as the $\mathbf{CC_p}$'s belief for the $\mathbf{S_i}$'s reliability. In realistic settings, the $\mathbf{CC_p}$ will have to face the type uncertainty. However, the possibility of repeated interaction can provide the $\mathbf{CC_p}$ with the ability to learn, progressively updating his beliefs about the type of $\mathbf{S_i}$. In the proposed algorithm, reinforcement learning mechanism gives the $\mathbf{CC_p}$ the opportunity, through observation of the outcome of team actions, to update his beliefs about the types of team members. Belief updates using the proposed reinforcement learning model will in turn influence future team formation decisions, which will be taken in a manner that is sequentially rational.

In order to make the proposed reinforcement learning model apply to realistic circumstances, we assume only limited observability of the realized outcomes: the CC_p only observes the outcome of team's actions, and the process then repeats. In this study, we cast the team formation problem as a learning-based repeated stochastic game model. Let the CC_p have belief vector $\mathbb{B} = [\mathscr{B}_{S_1} \ldots \mathscr{B}_{S_i} \ldots \mathscr{B}_{S_n}]$ about the types of all SN users where $0 \leq \mathscr{B}_{S_i} \leq 1$ represents the belief of S_i's influence power; $\mathscr{B}_{S_i}$ is measured as the S_i's closeness to his locally connected SN users. Therefore, the value of $\mathscr{B}_{S_i}$ is the expected value of $(\Lambda_{S_i}/\mathbb{R}_{S_i})$. At first, the CC_p doesn't know each user's B information, but can learn it based on the reinforcement learning model.

In the reinforcement learning method, the basic idea of value is an expected reward value of all future strategies, and the learner updates its value. Traditionally, the learner considers the expected future value with a discount factor [1]. However, unlike typical reinforcement learning equations, we focus on the recent outcome and past histories of individual influence powers of SN users. Due to the SN's uncertainty and computational complexity, it is hard to consider all outcomes when computing the value of any possible team formation. Therefore, the optimal approach for the team formation problem would be impossible. In the proposed scheme, the aim is to abstract away from the optimal team formation process; we deal only with current belief of $\mathbb{B}$ while considering the resulting team outcome. After the team formation at the time t, the S_i's belief at time $t + 1(\mathscr{B}_{S_i}^{t+1})$ is dynamically adjusted as follows:

$$\mathscr{B}_{S_i}^{t+1} = \begin{cases} \mathscr{B}_{S_i}^t, & \text{if } S_i \notin \mathbb{T}_t \\ \min\{\max\{[\mathscr{B}_{S_i}^t + (\eta \times (\hat{\mathscr{B}} - \mathscr{B}_{S_i}^t))], 0\}, 1\}, & \text{if } S_i \in \mathbb{T}_t, \end{cases} \tag{13.2}$$

$$\text{s.t.,} \quad \hat{\mathscr{B}} = \left(\frac{1}{|\mathbb{T}_t|} \times \left(\sum_{S_k \in \mathbb{T}_k} \left(\frac{\mathscr{D}_{S_i}(\mathscr{A}_{S_i}^t)}{\mathscr{T}_{S_i}(\mathscr{A}_{S_i}^t)} \right) \right) \right)$$

where $\mathbb{T}_t$ is the formed team at the time t, and η is the learning rate, which models the rate of updating value. $\mathscr{D}_{S_i}(\mathscr{A}_{S_i}^t)$ or $\mathscr{T}_{S_i}(\mathscr{A}_{S_i}^t)$ the S_i's real outcome (or expected outcome) due to the team formation $\mathbb{T}_t$. It is estimated based on the S_i's friends, i.e., directly connected S_i's neighbors ($\mathbb{R}_{S_i}$), and S_i's friends of friends, i.e., loosely connected SN users to the S_i. By considering the trickle down effect, $\mathscr{D}_{S_i}(\mathscr{A}_{S_i}^t)$ and $\mathscr{T}_{S_i}(\mathscr{A}_{S_i}^t)$ are defined as follows:

$$\begin{cases} \mathscr{D}_{S_i}\left(\mathscr{A}_{S_i}^t\right) = \left[\mathscr{H}(S_i) \times \theta_{S_i} \times \log_{\gamma_x}\left(\exp\left(\frac{\mathscr{A}_{S_i}^t}{\max\limits_{S_k \in T_t}\{\mathscr{A}_{S_k}^t\}} \right) \right) \right] \\ \quad + \sum_{S_r \in V, S_r \notin \mathbb{R}_{S_i}} \left((\mathscr{H}(S_i))^{\frac{1}{\min\{\rho(S_i,S_r)\}}} \times \theta_{S_i} \times \log_{\gamma_x}\left(\exp\left(\frac{\mathscr{A}_{S_i}^t}{\max\limits_{S_k \in T_t}\{\mathscr{A}_{S_k}^t\}} \right) \right) \right) \\ \mathscr{T}_{S_i}(\mathscr{A}_{S_i}^t) = \left[\mathscr{B}_{S_i}^t \times \theta_{S_i} \times \log_{\gamma_x}\left(\exp\left(\frac{\mathscr{A}_{S_i}^t}{\max\limits_{S_k \in T_t}\{\mathscr{A}_{S_k}^t\}} \right) \right) \right] \\ \quad + \sum_{S_r \in V, S_r \notin \mathbb{R}_{S_i}} \left((\mathscr{B}_{S_i}^t)^{\frac{1}{\min\{\rho(S_i,S_r)\}}} \times \theta_{S_i} \times \log_{\gamma_x}\left(\exp\left(\frac{\mathscr{A}_{S_i}^t}{\max\limits_{S_k \in T_t}\{\mathscr{A}_{S_k}^t\}} \right) \right) \right) \end{cases}$$

$$\tag{13.3}$$

where Υ_x is the control parameter for the requested application task, and $\min\{\rho(S_i, S_r)\}$ is the minimum number of social connections to reach the S_i. After each team formation time, the CC_p observes the subsequent observation of outcome and updates $\mathscr{B}$ values under the current obtained information. Based on the updated $\mathscr{B}$ values, we consider the scenario that the CC_p selects team members to maximize the payoff function as follows:

$$\max_{\mathbb{A}_t}\left(\mathbb{U}_t^{CC_p}(\mathbb{A}_t)\right) = \max_{S_k \in \mathbb{T}_t}\left\{\sum_{S_k \in \mathbb{T}_t}\mathscr{T}_{S_k}\left(\mathscr{A}_{S_k}^t\right)\right\} \qquad (13.4)$$

$$\text{s.t.,} \quad \sum_{S_k \in T_t}\mathscr{A}_{S_k}^t \leq \mathscr{W}$$

where $\mathbb{U}_t^{CC_p}(\mathbb{A}_t)$, $\mathbb{A}_t$, $\mathscr{A}_{S_k}^t$ are $\mathbb{U}^{CC_p}(\mathbb{A})$, $\mathbb{A}$, $\mathscr{A}_{S_k}$ values at the time t, respectively.

13.3.3 The Main Algorithm Steps of the Proposed Scheme

Influential maximization is a hot research topic in SNs. However, the current research has strongly focused on the diffusion process of "*word-of-mouth*" effect. In [8], Kempe et al. showed that the influential maximization problem is NP-Hard, and an effective solution for this problem is left as an open problem. In this study, SNS and CC technologies have captured to develop a novel SC system, and S. Kim offers a new influential maximization algorithm through the flexible and effective CC resource management. By adopting a reinforcement learning based one-player game model, we can detect the most influential users in SNs. To maximize the SC system performance, the type belief of each SN user is adaptively adjusted based on the dynamics of the interactive feedback process, and the team formation process will repeat at each time round. Therefore, the developed algorithm is formulated as a repeated game model; one decision might affect the next decisions in a step-by-step manner. This approach is realistic in real-world SC system operations. The proposed algorithm is described by the following seven major steps.

- **Step 1:** Control parameters $\mathbf{V}$, $\mathbf{E}$, n, m, $f_{\mathbf{V}}$, $\psi(\cdot)$, $\mathscr{W}$, η and Υ_x are given from the simulation scenario.
- **Step 2:** Initial time, i.e., $t = 0$, $\mathscr{B}$ values for SN users are equally distributed. This starting guess guarantees that each S enjoys the same benefit at the beginning of the game.
- **Step 3:** At each time round (t), some users in the SN randomly requests the CC services to execute their task applications, which have different Υ_x values and require different amounts of the CC resource.
- **Step 4:** According to (13.4), the CC_p dynamically selects some users to form a team $(\mathbb{T})$, and allocates the CC resource to maximize the social welfare.

- **Step 5:** Based on the current resource allocation, the CC_p adaptively adjusts the $\mathscr{B}$ values using (13.2) and (13.3). This reinforcement learning approach can predict the future influence power of each user based on the historical data.
- **Step 6:** During the step-by-step iteration, previous decisions are adaptively adjusted based on the dynamics of repeated game process.
- **Step 7:** Under the real-world SC environments, SN users and the CC_p are mutually dependent on each other to maximize the social welfare, and they constantly are self-monitoring the current SC system conditions; proceeds to **Step 3** for the next iteration.

13.4   Summary

In this work, S. Kim addresses a novel and challenging influential maximization problem in the SC system. Unlike the traditional methods, the proposed scheme sophisticatedly combines the SNS and CC technologies to maximize the social welfare. According to the basic idea of repeated game model and reinforcement learning algorithm, we can effectively select the most influential users while maximizing the CS system performance. Based on the iterative feedback process, all control decisions are dynamically adjusted. Under diverse SC system environments, the proposed repeated game based approach is a more realistic methodology for finding an effective solution with practical assumptions. For the future research, the open issues and practical challenges are data mining, security, social bargaining, and reinforcement learning in SC system operations. In particular, S. Kim plans to investigate the social influence mining algorithm in SNs. The combination of social influence mining and influence maximization will be a key issue that enables a prevalent online marketing in social media. In addition, a higher resolution multi-period scheme with inter-temporal constraints can be a potential direction and another possible extension to this work. Another future direction is to look for hybrid approaches that combine the advantages of different reinforcement algorithms to further improve the efficiency and effectiveness of influence maximization problem.

References

1. Chalkiadakis, G.: A bayesian approach to multiagent reinforcement learning and coalition formation under uncertainty. Doctoral Dissertation, Toronto (2007)
2. Chard, K., Bubendorfer, K., Caton, S., Rana, O.F.: Social cloud computing: a vision for socially motivated resource sharing. Services computing. IEEE Trans. Serv. Comput. (2012). https://doi.org/10.1109/TSC.2011.39
3. Chen, W., Wang, Y., Yang, S.: Efficient influence maximization in social networks. In: ACM KDD'2009 (2009). https://doi.org/10.1145/1557019.1557047

4. Deng, X., Pan, Y., Wu, Y., Gui, J.: Credit Distribution and influence maximization in online social networks using node features. In: IEEE FSKD'2015 (2015). https://doi.org/10.1109/FSKD.2015.7382274

5. Du, Z., Wang, Q., Fu, X., Liu, Q.: Integrated and flexible data management for cloud social network service platform on campus. In: IEEE ICCSNT'2012 (2012). https://doi.org/10.1109/ICCSNT.2012.6526148

6. Hwang, Y.C., Shiau, W.C.: Exploring imagery-driven service framework on social network service. In: IEEE/ACM ASONAM'2012 (2012). https://doi.org/10.1109/ASONAM.2012.192

7. Jiao, Y., Wang, Y., Yuan L., Li, L.: Cloud and SNS supported collaboration in AEC industry. In: IEEE CSCWD'2012 (2012). https://doi.org/10.1109/CSCWD.2012.6221919

8. Kempe, D., Kleinberg, J., Tardos, É.: Maximizing the spread of influence through a social network. In: ACM SIGKDD'2003 (2003). https://doi.org/10.1145/956750.95676

9. Kim S.: Game Theory Applications in Network Design. IGI Global, Hershey (2014)

10. Kim, S.: Dynamic social cloud management scheme based on transformable Stackelberg game. EURASIP J. Wirel. Commun. Netw. (2016). https://doi.org/10.1186/s13638-016-0543

11. Kim, J., Kim, S., Yu, H.: Scalable and parallelizable processing of influence maximization for large-scale social networks. In: IEEE ICDE'2013 (2013). https://doi.org/10.1109/ICDE.2013.6544831

12. Lee, K., Shin, I.: User mobility model based computation offloading decision for mobile cloud. J. Comput. Sci. Eng. (2015). https://doi.org/10.5626/JCSE.2015.9.3.155

13. Liu, Y., Sun, Y., Ryoo, J., Rizvi S., Vasilakos, A.V.: A survey of security and privacy challenges in cloud computing: solutions and future directions. J. Comput. Sci. Eng. (2015). https://doi.org/10.5626/JCSE.2015.9.3.119

14. Mihara, S., Tsugawa, S., Ohsaki, H.: Influence maximization problem for unknown social networks. In: IEEE/ACM ASONAM'2015 (2015). https://doi.org/10.1145/2808797.2808885

15. Pan, Y., Hu, N.: Research on dependability of cloud computing systems. In: IEEE ICRMS'2014 (2014). https://doi.org/10.1109/ICRMS.2014.7107234

16. Tang, C., Li, Q., Xiong, Y., Wen, S., Liu, A., Zhong, F.: Dominance-based service selection scheme with concurrent requests. J. Comput. Sci. Eng. (2012). https://doi.org/10.5626/JCSE.2012.6.2.89

17. Zhu, W., Lee, C.: A new approach to web data mining based on cloud computing. J. Comput. Sci. Eng. (2014). https://doi.org/10.5626/JCSE.2014.8.4.181

Part III
Game Theory in Smart Grid

Chapter 14
Noncooperative Energy Charging and Discharging Game for Smart Grid

Hung Khanh Nguyen and Ju Bin Song

14.1 Introduction

Since electric vehicles[1] will be widely used in future transportation systems, a significant new load requires on the existing energy distribution system. Thus, if the charging process for a large number of electric vehicles is not coordinated, it can easily overload the grid capacity at peak hours and endanger the safe operation of the smart grid system [6, 9, 18, 23]. In the United States, the average car is driven for approximately one hour a day, and for the rest of the day is parked, thus spending most of its time in a garage and, in the case of PHEVs, connected to the smart grid [10]. This provides great opportunities for the building that houses the garage as the PHEV batteries can either serve as a distributed energy storage resource or add to the load [25]. Therefore, by properly charging and discharging their batteries, PHEVs can help to smooth a building's energy consumption profile and reduce its energy cost. A smooth building energy consumption profile, which includes peak clipping, valley filling, load shifting, and flexible load shape [5], is one of the key design objectives of the demand side management in the future smart grid [4, 13, 14]. Moreover, by discharging a vehicle's battery into a building's charging station, which is defined as a vehicle-to-building (V2B) operation [17], we can increase the flexibility and reliability of the electrical distribution operation. V2B operation will provide extra benefits to the vehicle owners, and reduce the building energy

---

[1]Electric vehicles and PHEVs (Plug-in Hybrid Electric Vehicles) are interchangeable.

H. K. Nguyen · J. B. Song (✉)
Department of Electronic Engineering, Kyung Hee University, Yongin, Gyeonggi, South Korea
e-mail: jsong@khu.ac.kr

© Springer Nature Switzerland AG 2019
J. B. Song et al. (eds.), *Game Theory for Networking Applications*,
EAI/Springer Innovations in Communication and Computing,
https://doi.org/10.1007/978-3-319-93058-9_14

cost based on the demand side management program. This motivated us to develop an effective charging and discharging algorithm for multiple PHEV batteries in a smart building to optimize the energy consumption profile.

In this chapter, we introduce a noncooperative game theoretical framework [8] for the charging and discharging of multiple PHEV batteries to optimize the energy load profile of a smart grid building, in which the players are the PHEVs and their strategies are charging and discharging profiles, that we proposed in [15]:

- First, we design a centralized charging and discharging problem for multiple PHEV batteries in a smart building to reduce the peak demand, which minimizes the Square Euclidean Distance (SED) between the instantaneous load demand and the average demand, called the SED minimization problem.
- Second, we design an energy cost sharing model and propose a distributed algorithm to encourage PHEV owners to participate in the charging and discharging process.

The rest of this chapter is organized as follows. We describe the current state of the art of this research area in Sect. 14.2. The system model is introduced in Sect. 14.3. In Sect. 14.4, we analyze noncooperative energy charging and discharging game model for the distributed design. We present our conclusions in Sect. 14.5.

14.2 Related Studies

In recent smart grid literature, a great number of papers report studies of charging schedules for PHEV batteries. In [7], the authors studied the charging sequence control problem for an electric vehicle in order to maximize its revenue in a given charging period under some design constraints, such as the energy restriction of its battery. By applying a dynamic programming technique, they derived the optimal charging sequence that would maximize the profit while satisfying the state-of-charge level required at the end of the charging period. In [2], the coordinated PHEV charging problem for minimizing power loss and voltage deviations was studied. The power losses minimization problem was formulated as a nonlinear minimization problem, which could be addressed as a sequential quadratic optimization. They proposed an algorithm for coordinated charging of PHEV batteries when the daily load profile is deterministic. When the historical data for the daily load profile were not available, the authors applied a stochastic programming technique to find the optimal charging profile for PHEVs. They also analyzed the optimal PHEV charging coordination using dynamic programming techniques and studied its impact on the distribution grid. Deilami et al. [3] studied a novel load management solution for coordinating the charging of multiple PHEVs in a smart grid system to minimize power loss and improve voltage profile. A real-time smart load management algorithm to coordinate multiple PHEV batteries was proposed to reduce the energy generation cost by incorporating time-varying market energy prices. Moreover, the algorithm also enables owners to start charging their PHEV as soon as possible considering priority-charging time zones while satisfying network operation criteria, such as power loss, power generation limits, and voltage profile.

One of the greatest advantages of the wide deployment of PHEVs is that their batteries can be served as distributed energy storage resources in the smart grid. Most of the time vehicles are idle at homes, or in parking lots, or garages; hence the vehicle's battery can constitute either part of the load or a generator. Moreover, the time during which electric vehicles are in parking lots is typically longer than the time required to charge them, which provides an opportunity to implement vehicle-to-grid (V2G) services [2, 24]. In [16], the authors studied an autonomous distributed V2G control scheme that included charging request, battery condition, and contribution to the smart grid. In [22], a fuzzy logic control technique was applied to design a V2G infrastructure. Two controllers were implemented, at the charging station and at the distribution node. The objective of a V2G controller is to control the power flow between a particular node and the charging station to meet peak power demand and reduce voltage sag. A model of an electric vehicle storage system integrated with a standard power system was investigated in [11]. The authors provided a decision-making strategy for the owners of electric vehicles to determine how to utilize the stored energy effectively by controlling the charging and discharging process, taking into consideration the vehicle battery's characteristics and, state of charge, the vehicle user's driving habits, and electricity prices. In [20], two algorithms to address the optimal charging control problem for PHEVs in deregulated electricity markets were proposed. The optimal solution that allowed the vehicle owner to achieve the minimum cost was obtained using a dynamic programming technique. The first algorithm optimizes the charging time and energy flow to reduce daily electricity costs without increasing battery degradation, and the second takes into consideration the vehicle's contribution to the grid support by allowing PHEVs to inject energy back into the grid, which additionally benefits to the PHEV owners.

Game theory has been applied to V2G as well as demand side management problems. In [28], the authors proposed a novel model of interaction between electric vehicles and aggregators in a V2G market in which electric vehicles participate in providing a frequency regulation service to the grid. A smart pricing model was introduced and a game theoretical approach was applied to a distributed design in which the players were the electric vehicles and their strategies were the control decisions on energy charging or discharging. The authors also showed that the distributed system obtains the same performance as a centralized controlled system. The non-cooperative game approach was applied to model the competitive situation between a number of PHEV groups in an energy trading scenario involving PHEVs and distribution grids in [21]. Each PHEV group determines the maximum amount of energy surplus to sell in order to maximize a utility function.

Therefore, in [15], we applied a non-cooperative game approach to model the interaction between the charging and discharging process of multiple PHEV batteries resulting in reducing the peak demand in a smart building, and provide a distributed algorithm in which each PHEV battery tries to minimize the energy charging cost.

14.3 System Model

We consider a smart building with a charging station for a set $\mathcal{N}$ of $N \triangleq |\mathcal{N}|$ PHEVs in [15]. The charging station is assumed to be equipped with a bidirectional charger and controller. Moreover, the charging station can allow the electric vehicle batteries in either charge or discharge mode. This assumption is reasonable in the future smart grid due to the recent advancements in smart grid technologies [12, 26]. The charging time horizon is divided into a set $\mathcal{T}$ of $T \triangleq |\mathcal{T}|$ equal length time slots. We assume that the building has an energy consumption profile over T time slots

$$l = [l^1, \ldots, l^t, \ldots, l^T]. \tag{14.1}$$

For each user (PHEV) i, we define an energy charging and discharging vector as

$$x_i = [x_i^1, \ldots, x_i^t, \ldots, x_i^T], \tag{14.2}$$

where x_i^t is the amount of energy user i uses to charge or discharge its battery at time slot t. It should be noted that, in this paper, we use boldface letters to denote vectors. The user i is charging when $x_i^t > 0$, discharging when $x_i^t < 0$, and idle when $x_i^t = 0$. We restrict the minimum and maximum energy charge/discharge for each user i in a time slot

$$-x_i^{\max} \leq x_i^t \leq x_i^{\max}, \tag{14.3}$$

where $x_i^{\max}$ is the maximum charging/discharging of user i.

Let x_i^0 be the battery level when user i begins to charge. When charging is complete, each user wants its battery to have a predetermined energy target level B_i. Therefore, the energy demand that user i needs for charging its battery can be calculated as

$$E_i = B_i - x_i^0. \tag{14.4}$$

Then, the constraint for total energy demand of user i is

$$\sum_{t=1}^{T} x_i^t = E_i. \tag{14.5}$$

The constraint (14.5) means that the total energy consumption of user i over T time slots must be equal to the total energy demand to reach the target level for its battery.

For each user, the charging and discharging schedule also depends on the scheduling plan in the previous time slots. At each time slot, after charging or discharging, the battery must not be over-charged or discharged. Therefore, we add the constraints:

$$0 \leq x_i^0 + \sum_{k=1}^{t} x_i^k \leq C_i, \forall t \in \mathscr{T}, \tag{14.6}$$

where C_i is the battery capacity of user i.

For each time slot t, the total energy consumption of a building cannot exceed the maximum allowable load, to ensure safety operation, and cannot be less than zero

$$0 \leq l^t + \sum_{i=1}^{N} x_i^t \leq L_{\max}, \tag{14.7}$$

where $L_{\max}$ is the maximum allowable load at each time slot. The first inequality in the constraint (14.7) ensures that the building cannot provide power back to the grid.

For each user i, we can define a *feasible* energy charging and discharging set as

$$\mathscr{X}_i = \{x_i \mid \text{constraints (14.3), (14.5), (14.6), (14.7)}\}. \tag{14.8}$$

Using the feasible set of energy charging and discharging vectors for each user, we define the energy charging and discharging optimization problem for the smart grid building in the next section.

14.4 Noncooperative Energy Charging and Discharging Game Model

14.4.1 Centralized Problem

Firstly, we consider a centralized control system where the central planner schedules the charging and discharging process of users to achieve the target system performance. From the view-point of the building controller, the load profile should be as constant as possible, taking into account the extra energy demand of the vehicle's battery charging. Any energy demand less than the average demand will cause poor utilization of the existing infrastructure system and any energy demand exceeding the average demand will increase the energy cost as well as endanger the reliability of the building's operation. Therefore, we formulate a centralized optimization problem, the SED minimization problem, for the optimal charging and discharging of multiple PHEV batteries. The central controller searches energy charging and discharging schedules that minimize the square Euclidean distance between the instantaneous load profile and average demand.

Let E_{avg} denote the average energy demand of the smart building over T time slots

$$E_{avg} = \frac{\sum_{t=1}^{T} l^t + \sum_{i=1}^{N} E_i}{T}. \tag{14.9}$$

The square Euclidean distance between the instantaneous target load demand and average demand can be calculated as

$$\text{SED} \triangleq \sum_{t=1}^{T} \left(l^t + \sum_{i=1}^{N} x_i^t - E_{avg} \right)^2. \tag{14.10}$$

Then, the SED minimization problem can be formulated as

$$\min_{\forall i,\, x_i \in \mathscr{X}_i} \sum_{t=1}^{T} \left(l^t + \sum_{i=1}^{N} x_i^t - E_{avg} \right)^2. \tag{14.11}$$

Theorem 1 *The optimization problem* (14.11) *is convex and thus has a unique optimal solution.*

Proof Since for each user $i \in \mathscr{N}$, the feasible set $\mathscr{X}_i$ contains only linear constraints, then it is convex and also compact. Since the objective function is quadratic, it is a strictly convex function. Therefore, the optimization problem (14.11) is convex and has a unique optimal solution.   □

The optimal solution of the optimization problem (14.11) can be obtained in a centralized fashion using convex programming techniques such as the Interior Point Method [1]. Thus, after collecting all the information from users, the building controller solves the optimization problem (14.11) to obtain the optimal schedules for the users. This requires users to reveal private data to the building and makes the centralized system difficult to implement. To overcome this issue, in the next section we propose a decentralized system by applying a game approach.

14.4.2 Decentralized Design

The optimization problem defined in the previous section can be solved in a centralized fashion to obtain the optimal solutions for the SED minimization problem. However, this system requires a central planner to collect all the user information, such as energy demand, battery capacity, and initial battery level. This causes a breach of the owner's privacy and requires an exchange of much information with the building controller, which is not practical due to the enormous amount of signaling required for this purpose. Therefore, it is more advantageous to design a distributed algorithm in which users independently determine their energy charging and discharging schedules to achieve the best system performance. This fact motivated us to propose an alternative decentralized mechanism to achieve the desired performance. In the distributed design, the users exchange only their energy schedules with the building controller. Moreover, the SED minimization is a preferable design objective for reducing the peak demand of the building, but this is not a major concern for the PHEV owners. From their point of view, only scheduling their energy charging and discharging process in such a way that the total payment at the end of each day can be minimized is important. Therefore, in this section, we introduce an energy cost sharing model to address the energy charging and discharging problem in which each user's objective is to minimize its energy payment to the building.

14.4.2.1 Energy Cost Model

In this subsection, we define the energy cost model for the energy consumption of users. At time slot t, the total energy demand L_t of the building is calculated as

$$L_t = l^t + \sum_{i=1}^{N} x_i^t. \qquad (14.12)$$

We define a cost function $C(L_t)$ which is the cost of buying an L_t unit of energy

$$C(L_t) = \delta L_t^2, \qquad (14.13)$$

where δ is a positive coefficient. The cost function $C(L_t)$ is an increasingly and strictly convex function [27]. From (14.13), we see that when the total demand increases, the energy cost increases. Then, the total energy cost of the building over T time slots can be calculated as

$$C_{total} = \sum_{t=1}^{T} C(L_t). \qquad (14.14)$$

For each user $i \in \mathcal{N}$, the payment at the end of each day should reflect its total charging energy demand; it also depends on the total cost of the building's energy demand. Let κ_i denote the proportion of user i's energy demand of that of the building

$$\kappa_i = \frac{E_i}{\sum_{t=1}^{T} l^t + \sum_{j=1}^{N} E_j}. \qquad (14.15)$$

Then, we assume user i's payment is proportional to the building's total energy demand

$$C_i = \kappa_i \sum_{t=1}^{T} C(L_t). \tag{14.16}$$

From (14.16), we can see that the owner of user i's payment will depend on the proportion of user i's total energy demand and the total energy cost of the building. For example, if the total charging demand of user i is twice that of user j, then the owner of user i will be charged twice as much as user the owner of user j. The exact payment depends on the total cost, C_{total}, of the building, which is derived from the cost at each time slot t.

14.4.2.2   Utility Function

For each user $i \in \mathcal{N}$, we define the utility function as the negative total energy cost for charging its battery over T time slots. Since the total energy cost for each user's charging depends not only on its own energy demand but also on that of other users from (14.16), we can derive the utility function as

$$U_i(\boldsymbol{x}_i, \boldsymbol{x}_{-i}) = -C_i$$

$$= -\kappa_i \sum_{t=1}^{T} \delta \left(l^t + \sum_{i=1}^{N} x_i^t \right)^2. \tag{14.17}$$

where $\boldsymbol{x}_{-i} \triangleq [\boldsymbol{x}_1, \boldsymbol{x}_2, \ldots, \boldsymbol{x}_{i-1}, \boldsymbol{x}_{i+1}, \ldots, \boldsymbol{x}_N]$ is the energy consumption profile vector chosen by all other users except user i. Using the utility function of each user in (14.17), we apply a non-cooperative game for the optimal charging and discharging of multiple electric vehicle batteries in the next section.

14.4.2.3   Noncooperative Energy Charging and Discharging Game Model

In a decentralized V2B system, each electric vehicle (or user) is an independent decision maker. Therefore, each user $i \in \mathcal{N}$ independently determines the energy charging strategy to minimize its total energy payment. Then, we can define a *Noncooperative Energy Charging and Discharging* (NECD) game $G = \{\mathcal{N}, \{\mathcal{X}_i\}_{i \in \mathcal{N}}, \{U\}_{i \in \mathcal{N}}\}$ among end users, by its three components: (1) the *players*, that is, the users in the set $\mathcal{N}$; (2) the *strategy* of each player $i \in \mathcal{N}$, which corresponds to an energy charging and discharging profile, $\boldsymbol{x}_i \in \mathcal{X}_i$; and (3) the *utility function* U_i of any user $i \in \mathcal{N}$ as in (14.17).

Based on the definition of the payoffs and strategies in the NECD game, the users try to select charging and discharging profiles that will minimize their energy costs.

We now consider the best response strategy, which is the user's choice to maximize its own payoff function assuming that all other users' strategies are fixed. For the NECD game, we define the concept of best response as:

Definition 1 For each user $i \in \mathcal{N}$, the best response strategy x_i^* is

$$x_i^* \in \arg\max_{x_i \in \mathcal{X}_i} U_i. \tag{14.18}$$

Thus, for any user $i \in \mathcal{N}$, when the strategies of the other users x_{-i} are fixed, any best response strategy x_i^* is at least as good as every other strategy in $\mathcal{X}_i$

Definition 2 Consider the NECD game $G = \{\mathcal{N}, \{\mathcal{X}_i\}_{i \in \mathcal{N}}, \{U_i\}_{i \in \mathcal{N}}\}$. A vector of strategies x^* constitutes a *Nash equilibrium*, which is a state in which no player can improve its utility by unilaterally deviating from its equilibrium strategy, if and only if it satisfies the set of inequalities

$$U_i(x_i^*, x_{-i}^*) \geq U_i(x_i, x_{-i}^*), \ \forall x_i \in \mathcal{X}_i, \ \forall i \in \mathcal{N}. \tag{14.19}$$

Theorem 2 *There exists a unique Nash equilibrium for the NECD game [19].*

Proof Since for each user $i \in \mathcal{N}$, the payoff function U_i is strictly concave with respect to x_i and the strategy set $\mathcal{X}_i$ is convex and also compact. Therefore, the NECD game is a strictly concave N-person game. Then, the Nash equilibrium always exists based on the [19, Theorem 1] and is unique due to [19, Theorem 3]. □

In the following theorem, we demonstrate the effectiveness and optimality of the Nash equilibrium.

Theorem 3 *The unique Nash equilibrium of the NECD game is the optimal solution of the SED minimization problem* (14.11).

Proof We will show that the global optimal solution of the problem (14.11) forms a Nash equilibrium for the NECD game. Let $\{x_1^*, \ldots, x_N^*\}$ be the optimal solution of the problem (14.11). We also define

$$J^* \triangleq \sum_{t=1}^{T} \left(l^t + \sum_{i=1}^{N} x_i^{t*} - E_{avg} \right)^2. \tag{14.20}$$

We can express (14.20) with respect to variables x_i^* and x_{-i}^* as

$$J^* \triangleq \sum_{t=1}^{T} \left(l^t + x_i^{t*} + x_{-i}^{t*} - E_{avg} \right)^2, \tag{14.21}$$

where x_{-i}^{t*} denotes the total energy demand of all users except user i at time slot t at optimality

$$x_{-i}^{t*} \triangleq \sum_{j \neq i}^{N} x_j^{t*}.$$ (14.22)

Since J^* is the optimal value of the problem (14.11), we have the following inequality for any arbitrary x_i

$$J^* \leq \sum_{t=1}^{T} \left(l^t + x_i^t + \sum_{j \neq i}^{N} x_j^{t*} - E_{avg} \right)^2.$$ (14.23)

From (14.21) and (14.23), we have

$$\sum_{t=1}^{T} \left(l^t + x_i^{t*} + x_{-i}^{t*} - E_{avg} \right)^2 \leq \sum_{t=1}^{T} \left(l^t + x_i^t + x_{-i}^{t*} - E_{avg} \right)^2,$$ (14.24)

or

$$\sum_{t=1}^{T} \left[(l^t + x_i^{t*} + x_{-i}^{t*})^2 - 2E_{avg}(l^t + x_i^{t*} + x_{-i}^{t*}) + E_{avg}^2 \right]$$

$$\leq \sum_{t=1}^{T} \left[(l^t + x_i^t + x_{-i}^{t*})^2 - 2E_{avg}(l^t + x_i^t + x_{-i}^{t*}) + E_{avg}^2 \right].$$ (14.25)

Since the average demand E_{avg} is constant, and we also have

$$E_{avg} \sum_{t=1}^{T} \left(l^t + x_i^{t*} + x_{-i}^{t*} \right) = E_{avg} \sum_{t=1}^{T} \left(l^t + x_i^t + x_{-i}^{t*} \right)$$

$$= E_{avg} \left(\sum_{t=1}^{T} l^t + \sum_{i=1}^{N} E_i \right).$$ (14.26)

We can rewrite the inequality (14.25) as

$$\sum_{t=1}^{T} \left(l^t + x_i^{t*} + x_{-i}^{t*} \right)^2 \leq \sum_{t=1}^{T} \left(l^t + x_i^t + x_{-i}^{t*} \right)^2.$$ (14.27)

Multiplying both sides of the inequality (14.27) by $-\kappa_i \delta$ and expressing it as a utility function in (14.17), we obtain

$$U_i(x_i^*, x_{-i}^*) \geq U_i(x_i, x_{-i}^*).$$ (14.28)

From a comparison of (14.28) and Definition 2, we can conclude that the optimal solution $\{x_1^*, \ldots, x_N^*\}$ forms a Nash equilibrium for the NECD game. However, from Theorem 2, the NECD game has a unique Nash equilibrium. Therefore, the optimal solution of the SED problem (14.11) is equivalent to the Nash equilibrium of the NECD game.   □

From Theorems 2 and 3, we see that by offering them a cost sharing model, the owners of PHEVs have an incentive to participate in the energy charging and discharging schedule problem in order to reduce their energy cost, as well as participate indirectly in solving the centralized optimization problem. The PHEV owners would be willing to participate in scheduling their energy consumption in order to pay less.

Consider user $i \in \mathcal{N}$, given x_{-i}, and assume that all other uses fixed their energy consumption profiles according to x_{-i}. The user i's best response can be determined by solving the local optimization problem

$$\max_{x_i \in \mathcal{X}_i} U_i(x_i, x_{-i}). \tag{14.29}$$

The maximization problem (14.29) can be replaced by the minimization problem

$$\min_{x_i \in \mathcal{X}_i} \kappa_i \sum_{t=1}^{T} \delta \left(l^t + x_i^t + \sum_{j \neq i}^{N} x_j^t \right)^2. \tag{14.30}$$

Let us denote the total demand of the building, except user i at time slot t, as

$$L_{-i}^t = l^t + \sum_{j \neq i}^{N} x_j^t. \tag{14.31}$$

We can rewrite (14.30) with respect to only the local variables of user i as

$$\min_{x_i \in \mathcal{X}_i} \kappa_i \sum_{t=1}^{T} \delta \left(L_{-i}^t + x_i^t \right)^2. \tag{14.32}$$

It should be noted that problem (14.32) now has only local variables for user i. Therefore, user i can solve the optimization problem (14.32) in a distributed fashion whenever receiving the load profile vector L_{-i} from all other users. In [15], the optimal solution of the whole problem can be obtained by the distributed scheduling algorithm for charging and discharging multiple PHEVs based on the game setup. Whenever user i receives the signal giving the total load of the building from the building controller, it calculates the L_{-i} by subtracting its own x_i. Then user i obtains the value of L_{-i} and solves the local problem (14.32) to find the best response x_i. After receiving the best response, it checks whether the new one is different from the current one and updates it. The new value of its load profile x_i will be sent to the building controller. The building controller will update the new total energy demand profile. This procedure will be repeated until convergence.

Theorem 4 *If users update their energy consumption vectors asynchronously, i.e., the users $i, j \in \mathcal{N}$ do not update their energy consumption scheduling vectors at the same time, then starting from any initial condition point, the distributed Algorithm 1 in [15] will converge to the Nash equilibrium point of the NECD game which coincides with the optimal solution of the centralized problem, the SED minimization problem.*

Proof In [15], the best response for each user $i \in \mathcal{N}$ is equivalent to solving the optimization problem (14.32). Therefore, if users play the distributed algorithm to choose the best responses sequentially in an asynchronous fashion, their energy cost either decreases or remains unchanged when the users update their energy consumption schedule. Since the energy cost of any user $i \in \mathcal{N}$ is bounded below (e.g., the energy cost is always nonnegative, see (14.16), the convergence to some fixed point is evident. At the fixed point of Algorithm 1 in [15], no user can improve its payoff by deviating from the fixed point when choosing the best response. This indicates that the fixed point is the Nash equilibrium of the NECD Game among users. Moreover, from Theorem 3, we can conclude that the convergence point of the Algorithm 1 coincides with the optimal solution of the SED problem (14.11). □

Each user is required to send its energy charging and discharging scheduling vector to the building controller. Therefore, it is possible that users will incorrectly report their energy charging and discharging scheduling vectors if this misreporting helps to achieve more benefits. However, we will show that all users will be truthful in reporting their energy charging and discharging vectors to the building controller as the following theorem.

Theorem 5 *In the proposed NECD Game, by using Algorithm 1 in [15], no user or group of users benefits by misreporting their energy charging and discharging scheduling vectors $x_i, \forall i \in \mathcal{N}$. That is, each user $i \in \mathcal{N}$ will pay a higher energy cost if it announces its energy charging and discharging schedule, x_i, incorrectly.*

Proof Let $x_1^*, \dots, x_N^*$ denote the optimal solution of the SED minimization problem (14.11). It should be noted that $x_1^*, \dots, x_N^*$ is also the Nash equilibrium of the NECD Game. Denote by $\bar{x}_1, \dots, \bar{x}_N$ the Nash equilibrium of the NECD game when a nonempty set of users $\mathcal{M} \subseteq \mathcal{N}$ is untruthful. We will show that each user $i \in \mathcal{M}$ would pay more to the building for energy than truthful users. We denote the final utility of user $i \in \mathcal{M}$ by U_i^* and $\bar{U}_i$ when user i is truthful and untruthful, respectively. According to the NECD Game, we have

$$U_i^* = -\kappa_i \sum_{t=1}^{T} \delta \left(l^t + x_i^{t*} + \sum_{j \neq i}^{N} x_j^{t*} \right)^2 \qquad (14.33)$$

and

$$\bar{U}_i = -\kappa_i \sum_{t=1}^{T} \delta \left(l^t + \bar{x}_i^t + \sum_{j \neq i}^{N} \bar{x}_j^t \right)^2, \qquad (14.34)$$

where $\boldsymbol{x}_i^*$ and $\bar{\boldsymbol{x}}_i$ are the truthful and untruthful energy consumption schedules of user i, respectively. Then, we have

$$U_i^* - \bar{U}_i = \kappa_i \delta \left[\sum_{t=1}^{T} \left(l^t + \sum_{i=1}^{N} \bar{x}_i^t \right)^2 - \sum_{t=1}^{T} \left(l^t + \sum_{i=1}^{N} x_i^{t*} \right)^2 \right]. \qquad (14.35)$$

Since the total energy consumption of users is fixed for both the truthful and untruthful cases, we have

$$E_{avg} \sum_{t=1}^{T} \left(l^t + \sum_{i=1}^{N} \bar{x}_i^t \right) = E_{avg} \sum_{t=1}^{T} \left(l^t + \sum_{i=1}^{N} x_i^{t*} \right)$$

$$= E_{avg} \left(\sum_{t=1}^{T} l^t + \sum_{i=1}^{N} E_i \right). \qquad (14.36)$$

From (14.35) and (14.36), we have

$$\frac{U_i^* - \bar{U}_i}{\kappa_i \delta} = \sum_{t=1}^{T} \left[\left(l^t + \sum_{i=1}^{N} \bar{x}_i^t \right)^2 - 2E_{avg} \left(l^t + \sum_{i=1}^{N} \bar{x}_i^t \right) + E_{avg}^2 \right]$$

$$- \sum_{t=1}^{T} \left[\left(l^t + \sum_{i=1}^{N} x_i^{t*} \right)^2 - 2E_{avg} \left(l^t + \sum_{i=1}^{N} x_i^{t*} \right) + E_{avg}^2 \right]$$

$$= \sum_{t=1}^{T} \left(l^t + \sum_{i=1}^{N} \bar{x}_i^t - E_{avg} \right)^2 \qquad (14.37)$$

$$- \sum_{t=1}^{T} \left(l^t + \sum_{i=1}^{N} x_i^{t*} - E_{avg} \right)^2$$

$$\geq 0. \qquad (14.38)$$

where the inequality results from the fact that $\boldsymbol{x}_1^*, \ldots, \boldsymbol{x}_N^*$ is the optimal solution of the centralized problem (14.11).

From (14.37), we can obtain that $U_i^* \geq \bar{U}_i$. Therefore, user i does not benefit from untruthfully reporting its energy charging and discharging schedule. The best strategy of user i is to report its true optimal energy charging and discharging schedule. There is no incentive for users to cheat since this may lead to a loss in utility or an increased energy cost of the cheating users or the group of cheating users as well as other users in the system. Therefore, the distributed Algorithm 1 [15] is strategy-proof. □

From Theorem 5, we can guarantee that our proposed distributed algorithm will achieve the Nash equilibrium point of the NECD game, which is also the optimal solution of the centralized optimization problem (14.11).

14.5 Conclusion

In this chapter, we have formulated a noncooperative charging and discharging game problem for smart grid. We have designed a cost sharing model in distributed manner to achieve the Nash equilibrium, in which each PHEV tries to minimize its energy charging cost. The game framework can reduce the peak energy demand of smart grid building and the total energy cost. This game framework can be extended in many ways. First, a pricing model should be designed for the V2B operation mode, which will give greater incentive to PHEV owners to participate in the discharging of energy back to the building. Second, the discharging process should take into account the impacts on battery life. Third, we can consider this problem under the assumption that the building is equipped with a renewable energy generator. By using the PHEV batteries as energy storage systems, the building can effectively reduce the total energy cost as well as the peak power demand. Finally, we can apply stochastic optimization techniques to consider this problem under some uncertainties, such as the power price and the load profile of the building.

References

1. Boyd, S., Vandenberghe, L.: Convex Optimization. Cambridge University Press, New York (2004)
2. Clement-Nyns, K., Haesen, E., Driesen, J.: The impact of charging plug-in hybrid electric vehicles on a residential distribution grid. IEEE Trans. Power Syst. **25**, 371–380 (2010)
3. Deilami, S., Masoum, A.S., Moses, P.S., Masoum, M.A.S.: Real-time coordination of plug-in electric vehicle charging in smart grids to minimize power losses and improve voltage profile. IEEE Trans. Smart Grid **2**, 456–467 (2011)
4. Fang, X., Misra, S., Xue, G., Yang, D.: Smart grid - the new and improved power grid: a survey. IEEE Commun. Surv. Tutorials **14**, 944–980 (2011)
5. Gellings, C.W.: The concept of demand-side management for electric utilities. Proc. IEEE **73**, 1468–1470 (1985)
6. Hadley, S.W., Tsvetkova, A.A.: Potential impacts of plug-in hybrid electric vehicles on regional power generation. Electr. J. **22**, 25–37 (2008)
7. Han, S., Han, S., Sezaki, K.: Development of an optimal vehicle-to-grid aggregator for frequency regulation. IEEE Trans. Smart Grid **1**, 65–72 (2010)

8. Han, Z., Niyato, D., Saad, W., Basar, T., Hjorungnes, A.: Game Theory in Wireless and Communication Networks: Theory, Models and Applications. Cambridge University Press, New York (2011)
9. Hossain, E., Han, Z., Poor, H.V.: Smart Grid Communications and Networking. Cambridge University Press, New York (2012)
10. Kempton, W., Udo, V., Huber, K., Komara, K., Letendre, S., Baker, S., Brunner, D., Pearre, N.: A test of vehicle-to-grid (V2G) for energy storage and frequency regulation in the PJM system. System **2008**, 32 (2009)
11. Ma, Y., Houghton, T., Cruden, A., Infield, D.: Modeling the benefits of vehicle-to-grid technology to a power system. IEEE Trans. Power Syst. **27**, 1012–1020 (2012)
12. Massoud Amin, S., Wollenberg, B.F.: Toward a smart grid: power delivery for the 21st century. IEEE Power Energ. Mag. **3**, 34–41(2005)
13. Mohsenian-Rad, A.H., Leon-Garcia, A.: Optimal residential load control with price prediction in real-time electricity pricing environments. IEEE Trans. Smart Grid **1**, 120–133 (2010)
14. Mohsenian-Rad, A.H., Wong, V.W.S., Jatskevich, J., Schober, R., Leon-Garcia, A.: Autonomous demand-side management based on game-theoretic energy consumption scheduling for the future smart grid. IEEE Trans. Smart Grid **1**, 320–331 (2010)
15. Nguyen, H.K., Song, J.B.: Optimal charging and discharging for multiple PHEVs with demand side management in vehicle-to-building. J. Comun. Netw. **14**, 662–671 (2012)
16. Ota, Y., Taniguchi, H., Nakajima, T., Liyanage, K.M., Baba, J., Yokoyama, A.: Autonomous distributed V2G (vehicle-to-grid) considering charging request and battery condition. In: Proceedings of Innovative Smart Grid Technologies Conference Europe (ISGT Europe) (2010)
17. Pang, C., Dutta, P., Kezunovic, M.: BEVs/PHEVs as dispersed energy storage for V2B uses in the smart grid. IEEE Trans. Smart Grid **3**, 473–482 (2012)
18. Roe, C., Meisel, J., Meliopoulos, A.P., Evangelos, F., Overbye, T.: Power system level impacts of PHEVs. In: Proceedings of Hawaii International Conference on System Sciences, Hawaii (2009)
19. Rosen, J.B.: Existence and uniqueness of equilibrium points for concave N-Person games. Econometrica **33**, 520–534 (1995)
20. Rotering, N., Ilic, M.: Optimal charge control of plug-in hybrid electric vehicles in deregulated electricity markets. IEEE Trans. Power Syst. **21**, 1–9 (2010)
21. Saad, W., Han, Z., Poor, H.V., Basar, T.: A noncooperative game for double auction-based energy trading between PHEVs and distribution grids. In: IEEE International Conference on Proceedings of Smart Grid Communications (SmartGridComm) (2011)
22. Singh, M., Kumar, P., Kar, I.: Implementation of vehicle to grid infrastructure using fuzzy logic controller. IEEE Trans. Smart Grid **3**, 565–577 (2012)
23. Sortomme, E., Hindi, M.M., MacPherson, S.D.J., Venkata, S.S.: Coordinated charging of plug-in hybrid electric vehicles to minimize distribution system losses. IEEE Trans. Smart Grid **2**, 198–205 (2011)
24. Tomic, J., Kempton, W.: Vehicle-to-grid power fundamentals: calculating capacity and net revenue. J. Power Sources **144**, 268–279 (2005)
25. Tomic, J., Kempton, W.: Using fleets of electric-drive vehicles for grid support. J. Power Sources **168**, 459–468 (2007)
26. Vojdani, A.: Smart integration: calculating capacity and net revenue. IEEE Power Energ. Mag. **6**, 71–79 (2008)
27. Wu, C., Mohsenian-Rad, H., Huang, J., Wang, A.Y.: Demand side management for wind power integration in microgrid using dynamic potential game theory. In: Proceedings of GLOBECOM Workshops (GC Wkshps). IEEE, Piscataway (2011)
28. Wu, C., Mohsenian-Rad, H., Huang, J.: Vehicle-to-aggregator interaction game. IEEE Trans. Smart Grid **3**, 434–442 (2012)

Chapter 15
Stackelberg Differential Game Based Charging Control of Electric Vehicles in Smart Grid

Haitao Xu, Hung Khanh Nguyen, Xianwei Zhou, and Zhu Han

15.1 Introduction

With the rapidly increasing demand for electricity, lots of renewable energies have been introduced to smart grid to satisfy the increasing electricity demands and to solve the pollution problems. Although the renewable energy has been widely utilized in smart grid, their power supply is not stable enough compared to the conventional grid. Equipped with the plug-in connector compatible, the electric vehicles (EVs) can be charged or discharged with aggregations in power gird [1]. Charged electricity is stored in batteries of EVs. Then the EVs can be considered as power resources in smart grid. Integrated EVs into smart grid networks have been recognized as one essential way in reducing the emission of green-house gases [2].

As the electricity price of the conventional gird varies over the time, the aggregation needs to pay more at the peak hour for electricity transactions. Through considering the EVs as the power resources, the electricity transactions of the aggregation can be reduced, because instead of charging from the aggregation, the EVs can sell electricity back to the aggregation in a lower price to reduce the energy cost [3]. In this paper, we study the charging control problems of the EVs in smart grid to control the electricity transactions between the aggregation and the EVs. The main challenge of the charging control problems are the transaction price control and charing/discharging power control. The aggregation decides electricity price for transactions between the aggregation and the EVs. The EVs decide the charing/discharging power amount based on the price.

H. Xu (✉) · X. Zhou
University of Science and Technology Beijing, Beijing, China

H. K. Nguyen · Z. Han
University of Houston, Houston, TX, USA

© Springer Nature Switzerland AG 2019
J. B. Song et al. (eds.), *Game Theory for Networking Applications*,
EAI/Springer Innovations in Communication and Computing,
https://doi.org/10.1007/978-3-319-93058-9_15

Lots of works have been down recently to the charging control problems of electric vehicles in smart grid [4–6]. A framework for controlling the charging and discharging processes of plug-in electric vehicles (PEVs) via pricing strategies is investigated in [4]. The objective of the aggregation is to choose a pricing strategy for the PEVs to sell the energy. The PEVs can seek the equilibrium through the proposed iterative algorithm. In [5], the optimal charging/discharging problem is formulated as a mixed discrete programming problem, and a decentralized algorithm is proposed based on the iterative water-filling to solve the formulated problem. In [6], cyber insurance is used for PEVs in V2G systems, and a Markov decision process framework is used to formulate the energy cost optimization problem. Each PEV can make the optimal decision on the charge or discharge based on the proposed learning algorithm.

In this paper, we aim to propose a Stackelberg differential game model for the charging control problem in smart grid. We use differential equations to describe the dynamic characteristics of the batteries in the aggregation and EVs. The main contributions of this paper are as follows:

- A Stackelberg differential game model is constructed to solve the charging control problem in smart grid, which is a one-leader-many-followers Stackelberg game model, combining the differential game to describe the dynamic of the batteries. The aggregation acts as the leader of the game, while the EVs act as the followers.
- The optimal electricity price and charging power are given based on the equilibrium solutions of the differential games. Through controlling the electricity price by the aggregation, the EVs are encouraged to sell the bought electricity back to the aggregation to earn some profits and to low the energy cost. The aggregation can control their electricity price to maximize its payoff.

The remainder of the paper is organized as follows: Sect. 15.2 introduces the system model and formulates the Stackelberg differential game based charging control problem. Section 15.3 provides the equilibrium solutions for the game leader and followers. Numerical simulations are given in Sect. 15.4. Finally, we conclude the work in Sect. 15.5.

15.2   System Model and Problem Formulation

We consider a system with one aggregation and a finite set $\mathcal{N} \triangleq \{1, 2, \ldots, N\}$ EVs . The aggregation acts as the game leader, while the EVs act as followers. The aggregation decides the price $u_a(t)$ for buying or selling electricity. EV i can decide to charge or discharge $p_i(t)$ amount of energy during a continuous observation period $[0, T]$. Let $x_a(t)$ and $x_i(t)$ denote the battery level of the aggregation and EV i at time t, respectively, which are called the system state. We have the following differential equations to describe the dynamic variation of the state,

$$\frac{dx_a(t)}{dt} = -\sum_{i=1}^{N}\frac{\alpha_i}{\beta_i}p_i(t) + \varepsilon_a x_a(t), \tag{15.1}$$

$$\frac{dx_i(t)}{dt} = \frac{\alpha_i}{\beta_i}p_i(t) + \varepsilon_i x_i(t), \tag{15.2}$$

where α_i is the energy conversion efficient of EV i. β_i is the battery capacity of EV i. ε_a and ε_i are the system consumption rate of the aggregation and EV i, respectively, which are measured in units of electricity per time. The initial energy level of the aggregation is denoted by $x_a(0)$, which is equal to the capacity β_a of the aggregation. The initial state (battery) level of EV i is assumed to be $x_i(0)$.

For the aggregation, its cost function consists of three parts. Firstly, compared to buying electricity from the power grid, the aggregation controls the electricity price $u_i(t)$ to earn profit for trading with electric vehicles. Specifically, the profit can be calculated as follows:

$$U_{pro}^a(t) = [u_a(t) - \pi_a(t)]^2, \tag{15.3}$$

where $\pi_a(t)$ is the price for buying electricity from the power grid at time t. Secondly, the aggregation also tries to control the payoff for trading electricity with electric vehicles, which is denoted by

$$U_{cos}^a(t) = \sum_{i=1}^{N}u_a(t)p_i^2(t). \tag{15.4}$$

Thirdly, the aggregation aims to have enough energy available to cope with the energy demands, even to serve more vehicles. Let $\bar{x}_a$ denote the target energy level for the aggregation, whose payoff function can be calculated by

$$U_{eng}^a(t) = (x_a(t) - \bar{x}_a)^2. \tag{15.5}$$

Based on the above assumptions, the instantaneous payoff function of the aggregation at time t can be defined as follows:

$$U^a(t) = \eta_a[u_a(t) - \pi_a(t)]^2 + \omega_a\sum_{i=1}^{N}u_a(t)p_i^2(t) + \upsilon_a(x_a(t) - \bar{x}_a)^2 \tag{15.6}$$

where η_a, ω_a, and υ_a are positive weighted factors. Based on the payoff function, we can find that the instantaneous payoff mainly depends on the transactions price $u_a(t)$. The objective of the aggregation is to find the optimal electricity price $u_a^*(t)$ that can maximize its payoff function over time interval $[0, T]$

$$\max_{u_a(t)} L^a(t) = \max_{u_a(t)} \left\{ \int_0^T \left\{ \eta_a [u_a(t) - \pi_a(t)]^2 + \omega_a \sum_{i=1}^N u_a(t) p_i^2(t) \right. \right.$$

$$\left. \left. + \upsilon_a (x_a(t) - \bar{x}_a)^2 \right\} e^{-rt} dt \right\} \tag{15.7}$$

subject to (15.1). Here, r is the discount rate.

Next, we will discuss how electric vehicles control $p_i(t)$ to minimize their energy cost. Generally speaking, the energy cost of each EV mainly consists of two parts. One is the trading cost, and another one is the storage cost. For EV i, the trading cost is mainly dependent on the energy control variable $p_i(t)$ and the trading price $u_a(t)$. The instantaneous trading cost is defined as a linear quadratic form as follows:

$$U_{tra}^i(t) = u(t) p_i^2(t). \tag{15.8}$$

Each EV wants to sell electricity back to the aggregation to earn profits, under the premise of enough energy when leaving, which will cause an additional storage cost. It is guaranteed that the battery of EV i is no less than a threshold $\bar{x}_i$ during parking, to insure enough available energy when leaving the aggregation. Then the storage cost is mainly dependent on the battery threshold, and the instantaneous storage cost is defined as follows:

$$U_{sto}^i(t) = (x_i(t) - \bar{x}_i)^2. \tag{15.9}$$

Based on the above assumptions, the total instantaneous cost of EV i at time t is denoted by

$$U^i(t) = \omega_i u_a(t) p_i^2(t) + \upsilon_i (x_i(t) - \bar{x}_i)^2, \tag{15.10}$$

where ω_i and υ_i are positive weighted factors. We can observe that the total instantaneous cost depends on the charging/discharging power $p_i(t)$ adopted by EV i, as well as the battery state $x_i(t)$ according to the differential equation in (15.2). For EV i, the objective is to find the optimal power trading strategy $p_i^*(t)$ that can minimize its cost function over time interval $[0, T]$,

$$\min_{p_i(t)} L_i(t) = \min_{p_i(t)} \left\{ \int_0^T \left[\omega_i u_a(t) p_i^2(t) + \upsilon_i (x_i(t) - \bar{x}_i)^2 \right] e^{-rt} dt \right\} \tag{15.11}$$

subject to (15.2).

15.3 Game Analysis

In this section, we analyze the optimal electricity pricing and charging problems for the aggregation and EVs, respectively. As it is a Stackelberg differential game, we need to get the optimal solution for each EV first, then the aggregation can make a decision on the electricity price based on the charging control solutions. In the following subsections, we first discuss the optimal charging control problem for each EV in a finite time horizon $[0, T]$. Then, the optimal strategy of the leader (the aggregation) can be obtained based on each EV's solutions. All the optimization problems can be solved based on the dynamic programming [7–9].

15.3.1 Equilibrium Solutions of the EVs

We firstly discuss the optimal charging control problem for the EVs and get the open loop equilibrium solutions to the EVS based on the dynamic programming.

Definition 1 For EV i, the charging power strategy $p_i^*(t)$ is optimal if the following inequality holds for all feasible control $p_i(t) \neq p_i^*(t)$,

$$L_i(p_i^*(t), x_i^*(t), t) \leq L_i(p_i(t), x_i(t), t). \tag{15.12}$$

With the definition of optimal strategy for the EVs, the definition of the open-loop equilibrium and corresponding state trajectory are given as follows:

Definition 2 A set of controls $\{p_i^*(t)\}$ constitutes an open loop equilibrium to the problem in (15.11), and $x_i^*(t)$ is the corresponding state trajectory, if there exists a costate function $\Lambda_i(t)$ such that the following relations are satisfied:

$$p_i^*(t) = \arg\min_{p_i(t)} \left\{ U^i(t)e^{-rt} + \Lambda_i(t)\frac{dx_i(t)}{dt} \right\}, \tag{15.13}$$

$$\dot{\Lambda}_i(t) = -\frac{\partial\left[U^i(t)e^{-rt} + \Lambda_i(t)\frac{dx_i(t)}{dt} \right]}{\partial x_i(t)}. \tag{15.14}$$

where (15.14) is an adjoint equation to describe the dynamics of a costate variable. The costate function is a function which associates with the state variable $x_i(t)$.

In order to get the equilibrium solutions to the optimal problem in (15.11), we need to construct the Hamiltonian system for each EV, and solve the optimal problems based on the Pontryagin's maximum principle. Here, the equilibrium solutions for the EVs are the solutions of the differential game, and also are the Stackelberg equilibrium solutions for the followers. The Hamiltonian system of EV i is as follows.

Definition 3 The Hamiltonian system of EV i is given by the following equation:

$$H_i(t) = U^i(t)e^{-rt} + \Lambda_i(t)\frac{dx_i(t)}{dt}, \tag{15.15}$$

where $\Lambda_i(t)$ is the costate function given by (15.14). With the definition of the Hamiltonian function, the minimization of (15.11) is the corresponding minimization of the Hamiltonian function, which is defined as follows:

$$H_i^*(t) = \min_{p_i^*(t)} H_i(p_i(t), x_i(t), t), \tag{15.16}$$

Lemma 1 *The optimal power solutions to EV i is*

$$p_i^*(t) = \frac{-\alpha_i \Lambda_i(t)}{2\beta_i \omega_i u_a(t)} e^{rt}. \tag{15.17}$$

where $\Lambda_i(t)$ is given by (15.14).

Proof Taking (15.2) and (15.10) into the Hamiltonian system of EV i, we have

$$H_i(t) = \left[\omega_i u_a(t)p_i^2(t) + \upsilon_i(x_i(t) - \bar{x}_i)^2\right]e^{-rt} + \Lambda_i(t)\left[\frac{\alpha_i}{\beta_i}p_i(t) + \varepsilon_i x_i(t)\right]. \tag{15.18}$$

Performing the indicated maximization yields

$$\frac{\partial H_i(t)}{\partial p_i(t)} = 2\omega_i u_a(t)p_i(t)e^{-rt} + \Lambda_i(t)\frac{\alpha_i}{\beta_i}. \tag{15.19}$$

Then we have

$$p_i^*(t) = \frac{-\alpha_i \Lambda_i(t)}{2\beta_i \omega_i u_a(t)} e^{rt}. \tag{15.20}$$

Performing the indicated maximization, we can also obtain the differential equation for $\Lambda_i(t)$ as follows:

$$\dot{\Lambda}_i(t) = -\frac{\partial H_i(t)}{\partial x_i(t)} = -2\upsilon_i(x_i(t) - \bar{x}_i)e^{-rt} + \varepsilon_i \Lambda_i(t). \tag{15.21}$$

As we have obtained the optimal equilibrium in Lemma 1, take $p_i^*(t)$ into (15.2), we have

$$\frac{dx_i(t)}{dt} = \frac{-\alpha_i^2 \Lambda_i(t)}{2\beta_i^2 \omega_i u_a(t)} e^{rt} + \varepsilon_i x_i(t). \tag{15.22}$$

Solving the differential equations in (15.20) and (15.21), we can get the corresponding state trajectory and costate functions.

15.3.2 Equilibrium Solutions of the Aggregation

Similarly, we can obtain the equilibrium solutions to (15.7) for the aggregation based on the dynamic programming.

Definition 4 For the aggregation, the electricity price strategy $u_a^*(t)$ is optimal if the following inequality holds for all feasible control $u_a(t) \neq u_a^*(t)$,

$$L^a(u_a^*(t), x_a^*(t), t) \geq L^a(u_a(t), x_a(t), t). \tag{15.23}$$

Definition 5 A set of controls $\{u_a^*(t)\}$ constitutes an open loop equilibrium to the problem in (15.7), and $x_a^*(t)$ is the corresponding state trajectory, if there exist costate functions $\lambda_a(t)$ and $\lambda_i(t)$ such that the following relations are satisfied:

$$u_a^*(t) = \arg \min_{u_a(t)} H_a(t), \tag{15.24}$$

$$\dot{\lambda}_a(t) = -\frac{\partial H_a(t)}{\partial x_a(t)}, \tag{15.25}$$

$$\dot{\lambda}_i(t) = -\frac{\partial H_a(t)}{\partial \Lambda_i(t)}, \tag{15.26}$$

where the Hamiltonian function of the aggregation is given by

$$H_a(t) = \left\{ \eta_a[u_a(t) - \pi_a(t)]^2 + \omega_a \sum_{i=1}^{N} u_a(t) p_i^2(t) \right.$$

$$\left. + \upsilon_a(x_a(t) - \bar{x}_a)^2 \right\} e^{-rt} + \lambda_a(t) \left[-\sum_{i=1}^{N} \frac{\alpha_i}{\beta_i} p_i(t) \right. \tag{15.27}$$

$$\left. + \varepsilon_a x_a(t) \right] + \sum_{i=1}^{N} \lambda_i(t) \dot{\Lambda}_i(t),$$

where $\lambda_a(t)$ and $\lambda_i(t)$ are costate functions for the aggregation. From the above equation, we can observe that the Hamiltonian function of the aggregation is more complex compared to that of the EVs. Acting as the game leader, the aggregation should consider the strategies of the EVs before making a decision on the electricity price. Then the aggregation considers the dynamics of the costate function $\Lambda_i(t)$ of all the followers in the Hamiltonian function.

Lemma 2 *The optimal electricity price for the aggregation is given by*

$$u_a^*(t) = \pi_a(t) - \frac{\omega_a}{2\eta_a} \sum_{i=1}^{N} p_i^2(t). \tag{15.28}$$

Proof Performing the indicated maximization yields

$$\frac{\partial H_a(t)}{\partial u_a(t)} = \left\{ 2\eta_a[u_a(t) - \pi_a(t)] + \omega_a \sum_{i=1}^{N} p_i^2(t) \right\} e^{-rt}. \tag{15.29}$$

Then we have

$$u_a^*(t) = \pi_a(t) - \frac{\omega_a}{2\eta_a} \sum_{i=1}^{N} p_i^2(t). \tag{15.30}$$

Performing the indicated maximization of the Hamiltonian function, we can also obtain the following differential equations to the dynamics of costate functions $\lambda_a(t)$ and $\lambda_i(t)$, i.e.,

$$\dot{\lambda}_a(t) = -\frac{\partial H_a(t)}{\partial x_a(t)} = -2\upsilon_a(x_a(t) - \bar{x}_a)e^{-rt} - \lambda_a(t)\varepsilon_a, \tag{15.31}$$

$$\dot{\lambda}_i(t) = -\frac{\partial H_a(t)}{\partial \Lambda_i(t)} = \varepsilon_i \lambda_i(t). \tag{15.32}$$

15.4 Numerical Simulations

In this section, we evaluate the performance of the proposed Stackelberg differential game model with MATLAB. We consider a system with one aggregation and three EVs. The time interval for the simulations is set to [0, 24]. To simulate the equilibrium solutions of the EVs, we first set the price for the electricity transactions as 1.5 cents per unit electricity. The battery capacity for each EVs is assumed to be the same, and is set as 50 kWh. The energy conversion efficient of each EV is 0.7. The system consumption rate is 0.05. The discount rate is set as 0.25. The other weighted constant parameters settings are given in Table 15.1.

Table 15.1 Weighted constant parameters setting

Parameters	EV1	EV2	EV3	Aggregation
ω	20	12	16	2.5
υ	50	50	50	50
$\bar{x}$	25	30	20	400

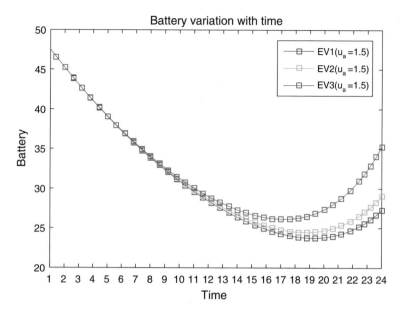

Fig. 15.1 Battery variation of the EVs with time

We first investigate the variation of the battery state for the EVs with initial state is 5, which means the EVs are fully charged at the beginning. The variation of the optimal state trajectory indicates the results of the charging control for each EV to minimize the energy cost. In Fig. 15.1, the battery level of each EV decreases at the beginning of the observation, which means the EVs act as the power resources to sell the electricity back to the aggregation. In this figure, the price for electricity transactions is 1.5 cents. The EVs want to sell the electricity to earn profits and to reduce their energy cost. When the battery level is less than the threshold, the EVs begin to buy the electricity to have enough energy when leaving.

For the aggregation, it controls the electricity price to maximize its payoff. Figure 15.2 shows the optimal electricity price variation of the aggregation with time. The electricity price of the power grid in this figure is set to be a time-varying parameter based on the price data form ComEd.com. We can observe that the electricity price of the power grid will affect the electricity price for transactions between the aggregation and EVs.

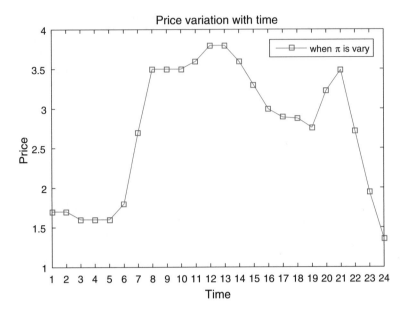

Fig. 15.2 Electricity price of the aggregation when the grid price is varying

15.5 Conclusion

In this paper, we use the Stackelberg differential game to formulate the charging control problem of the EVs in smart grid. In the Stackelberg differential game, the aggregation acts as the leader to determine the optimal electricity transactions price based on the equilibrium solutions. The EVs act as the followers to control their charging power amounts to earn profits from electricity transactions and to minimize the energy cost. The optimal solutions to the leader and followers can be obtained based on the open-loop equilibriums. Numerical analysis has been performed to show the correctness of the solutions. The impact of the leader's strategy on the solutions of the followers is investigated. We can also observe that the price of the power grid has effects on the price for transactions between the aggregation and the EVs.

Acknowledgements This work was supported by the National Science Foundation Project of China (No.61501026).

References

1. Couillet, R., Perlaza, S.M., Tembine, H., Debbah, M.: Electrical vehicles in the smart grid: a mean field game analysis. IEEE J. Sel. Areas Commun. **30**(6), 1086–1096 (2012)
2. Ma, Z., Callaway, D., Hiskens, I.: Decentralized charging control for large populations of plug-in electric vehicles: application of the Nash certainty equivalence principle. In: IEEE International Conference on Control Applications, Yokohama (2010)

3. Li, Y., Kaewpuang, R., Wang, P., Niyato, D., Han, Z.: An energy efficient solution: integrating plug-in hybrid electric vehicle in smart grid with renewable energy. In: Proceedings IEEE INFOCOM Workshops, Orlando (2012)
4. Gharesifard, B., Basar, T., Dominguez-Garcia, A.D.: Price-based distributed control for networked plug-in electric vehicles. In: American Control Conference, Washington (2013)
5. Xing, H., Fu, M., Lin, Z., Mou, Y.: Decentralized optimal scheduling for charging and discharging of plug-in electric vehicles in smart grids. IEEE Trans. Power Syst. **31**(5), 4118–4127 (2016)
6. Hoang, D.T., Wang, P., Niyato, D., Hossain, E.: Charging and discharging of plug-in electric vehicles (PEVs) in vehicle-to-grid (V2G) systems: a cyber insurance-based model. IEEE Access **5**, 732–754 (2017)
7. Basar, T., Olsder, G.J.: Dynamic Noncooperative Game Theory. Society for Industrial and Applied Mathematics, Philadelphia (1998)
8. Han, Z., Niyato, D., Saad, W., Basar, T., Hjorungnes, A.: Game Theory in Wireless and Communication Networks: Theory, Models and Applications. Cambridge University Press, Cambridge (2011)
9. Yeung, D.W.K., Petrosjan, L.A.: Cooperative Stochastic Differential Games. Springer, New York (2006)

Chapter 16
Day-Ahead Demand Management in Multi-Supplier Power Grid Under Transmission Constraints

Ivan V. Popov, Alexander Yu. Krylatov, Victor V. Zakharov, and Elena A. Lezhnina

16.1 Introduction

Traditionally, power grids have a central structure with a clear hierarchy. There are few power plants that produce and supply energy to a large area using transmission and distribution networks, and these power plants respond to a changing demand of consumers. However, due to the fast renewable energy development of recent decades, this situation is starting to change. New power grid architectures need to be created and studied in order to integrate smaller local renewable generators into the power grid while maintaining sustainability of the system.

In this paper, we formulate and consider a multi-supplier power grid model, where consumers need to conclude bilateral contracts with suppliers over a day-ahead period of time divided in several time slots (e.g., 24 h) [3, 6]. The distribution of flows in the network deserves special attention, since it is crucial for preventing overloads and other disturbances in transmission lines [4]. We describe consumers' costs as functions of their contract profiles, formulate a competitive game of consumers, and discuss possible schemes of demand response management for this model.

I. V. Popov
Control Methods and Robotics, TU Darmstadt, Darmstadt, Germany
e-mail: ipopov@rmr.tu-darmstadt.de

A. Y. Krylatov (✉)
Saint Petersburg State University, Saint-Petersburg, Russia
Solomenko Institute of Transport Problems, St. Petersburg, Russia
e-mail: a.krylatov@spbu.ru

V. V. Zakharov · E. A. Lezhnina
Saint Petersburg State University, Saint-Petersburg, Russia
e-mail: v.zaharov@spbu.ru

© Springer Nature Switzerland AG 2019
J. B. Song et al. (eds.), *Game Theory for Networking Applications*,
EAI/Springer Innovations in Communication and Computing,
https://doi.org/10.1007/978-3-319-93058-9_16

16.2   Model Description

A network is represented by a directed graph (V, A), where V is a set of nodes and A is a set of arcs. Let us enumerate nodes in V in the following manner: $V_Q = \{1, \ldots, m\}$ is a set of m energy consumers, $V_P = \{m + 1, \ldots, m + n\}$ is a set of n producers, and $V_O = \{m + n + 1, \ldots, |V|\}$ is a set of all other nodes.

In this work, we consider a day-ahead planning period divided into H intervals. Each consumer concludes bilateral energy purchase contracts with several producers for each time interval. By e_{ij}^h we denote an amount of energy to be delivered from producer $j \in V_P$ to consumer $i \in V_Q$ during the time interval $h \in \mathbf{H} = \{1, 2, \ldots, H\}$. We also use the following notation:

$$\mathbf{e}_i^h = \left(e_{i(m+1)}^h, \ldots, e_{i(m+n)}^h \right)^T \tag{16.1}$$

for a vector of i's contracts at a time interval h, and

$$\mathbf{E}_i = \left(\mathbf{e}_i^1, \ldots, \mathbf{e}_i^H \right) \tag{16.2}$$

for a matrix of all i's contracts.

Consumers need to meet their energy demands, both total for the whole day and minimal for each time interval $h \in \mathbf{H}$. We denote the total demand of consumer i by $D_i \geq 0$, and the minimal demand of the same consumer for a time interval h by $d_i^{\min}(h) \geq 0$. Therefore, we can write the demand constraints for $\mathbf{E}_i$:

$$\begin{aligned} \mathbf{1}_n^T \cdot \mathbf{e}_i^h &\geq d_i^{\min}(h), \\ \mathbf{1}_n^T \cdot \mathbf{E}_i \cdot \mathbf{1}_H &= D_i, \end{aligned} \tag{16.3}$$

where $\mathbf{1}_k = (1, 1, \ldots, 1)^T \in \mathbb{R}^k$.

Let us define energy balance b_k^h in a node $k \in V$ for a time interval $h \in \mathbf{H}$:

$$\begin{aligned} b_k^h &= -\mathbf{1}_n^T \cdot \mathbf{e}_k^h, \quad k \in V_Q, \\ b_k^h &= \sum_{i=1}^m e_{ik}^h, \quad k \in V_P, \\ b_k^h &= 0, \quad k \in V_O. \end{aligned} \tag{16.4}$$

This value reflects the amount of energy injected or withdrawn in a node during a specific time interval. It is negative for consumers and non-negative for producers, while we assume all other intermediate nodes to have zero energy balance.

Energy flows in a power grid are distributed according to Kirchhoff's laws, and we can find this distribution for a given set of energy balances and knowing parameters of grid links [1]. By $f_{kl}^h \geq 0$ we denote a flow in arc $(k, l) \in A$ at time interval h, and set $\mathbf{f}_h = \{f_{kl}^h, (k, l) \in A\}$ is a flow profile of all links at time h. We also denote by $\mathbf{f}_h(\mathbf{E}^h)$ the dependence of current flow from contract profile. This mapping is generally non-linear, and contract changes of a single consumer affect the flow distribution in the whole grid. Non-negativity of flows allows us to use traffic assignment algorithms for power load estimation [2, 4].

16.3 Game of Consumers

This section formulates and studies a consumer game as a model of interactions in the grid. First, we describe cost functions of consumers and formulate a game as a set of coupled cost minimization problems. In the second part of the section, the existence of Nash equilibria for the described game is discussed.

16.3.1 Consumer Cost Minimization

Each consumer tries to minimize their total costs over time span $\mathbf{H}$. These costs consist of two parts: generation costs and transmission costs. Generation costs can be assigned proportionally to the contracts between respective agents, while it is non-trivial to define the shares for use of transmission network.

More specifically, let $\alpha_j^h(b_j^h)$ denote a generation cost of a unit of energy at node $j \in V_P$ during time interval h. It is a function of total energy b_j^h to be generated at node j according to contracts with consumers $\mathbf{E}^h$. Hence, generation cost of consumer i during interval h can be determined in the following way:

$$G_i^h(\mathbf{E}^h) = \sum_{j=m+1}^{m+n} e_{ij}^h \cdot \alpha_j^h(b_j^h). \tag{16.5}$$

Transmission costs depend on the flow distribution $\mathbf{f}_h(\mathbf{E}^h)$. We define transmission cost for an arc $(k, l) \in A$ as a function $\beta_{kl}^h(f_{kl}^h)$ of the amount of flow using this arc. We call a set of functions $\Delta = \{\delta_{kl}^{i,h}(\mathbf{E}^h)\}$ a cost sharing rule, if it fulfills the following conditions:

$$\delta_{kl}^{i,h}(\mathbf{E}^h) \geq 0, \quad \forall (k, l) \in A, i \in V_Q, h \in \mathbf{H},$$

$$\sum_{i=1}^{m} \delta_{kl}^{i,h}(\mathbf{E}^h) = 1, \qquad \forall (k, l) \in A, h \in \mathbf{H}. \tag{16.6}$$

For a given cost sharing rule Δ the transmission cost of consumer i at interval h takes the form:

$$T_i^h(\mathbf{E}^h) = \sum_{(k,l) \in A} \delta_{kl}^{i,h}(\mathbf{E}^h) \cdot \beta_{kl}^h(f_{kl}^h(\mathbf{E}^h)). \tag{16.7}$$

Hence, the total cost of consumer i is

$$C_i(\mathbf{E}) = \sum_{h=1}^{H} \left(G_i^h(\mathbf{E}^h) + T_i^h(\mathbf{E}^h) \right), \tag{16.8}$$

where $\mathbf{E} = \{\mathbf{E}^1, \dots, \mathbf{E}^H\}$ is a total profile of all consumer contracts over the whole time span $\mathbf{H}$, and where the calculation of each transmission cost $T_i^h(\mathbf{E}^h)$ requires $\mathbf{f}_h(\mathbf{E}^h)$ for a respective time interval h.

We now formulate the game of consumers:

$$\underset{\mathbf{E}_i}{\text{minimize}} \qquad C_i(\mathbf{E}), \quad 1 \le i \le m, \tag{16.9}$$

$$\text{subject to} \qquad \mathbf{1}_n^T \cdot \mathbf{E}_i \cdot \mathbf{1}_H = D_i, \quad \forall i \in V_Q, \tag{16.10}$$

$$\mathbf{1}_n^T \cdot \mathbf{e}_i^h \ge d_i^{\min}(h), \quad \forall i \in V_Q, \forall h \in \mathbf{H}, \tag{16.11}$$

$$e_{ij}^h \ge 0, \qquad \forall i \in V_Q, \forall j \in V_P, \forall h \in \mathbf{H}. \tag{16.12}$$

In this game, contract matrix $\mathbf{E}_i$ is a strategy of consumer i. We denote by $\boldsymbol{\Sigma}_i$ a set of all i's feasible strategies, i.e., a set of all matrices $\{\mathbf{E}_i\}$ fulfilling the conditions (16.10)–(16.12).

16.3.2 Existence of Nash Equilibria

The idea of Nash equilibrium proved to be the most appropriate solution concept for competitive games. A set of agents' strategies is in Nash equilibrium, if none of agents may reduce their total cost by unilaterally changing their strategy. In our model, a total profile $\mathbf{E}^*$ is in Nash equilibrium, if the following conditions are fulfilled:

$$C_i(\mathbf{E}^*) \le C_i(\mathbf{E}_i, \mathbf{E}_{-i}^*), \forall \mathbf{E}_i \in \boldsymbol{\Sigma}_i, \tag{16.13}$$

where $\{\mathbf{E}_i, \mathbf{E}_{-i}^*\}$ is a total profile that differs from $\mathbf{E}^*$ only in component $\mathbf{E}_i$.

The existence of Nash equilibria in a consumer game strongly depends on the form of cost functions $\{\alpha_j^h(\cdot)\}$, $\{\beta_{kl}^h(\cdot)\}$ and the cost sharing rule Δ. Moreover, arguments of $\{\beta_{kl}^h(\cdot)\}$ are flows in the corresponding arcs. Hence, establishing the fact of equilibrium's existence is a non-trivial task.

Theorem 1 *Assume that a network contains no cycles, functions $\{\alpha_j^h(\cdot)\}$ are convex and increasing, functions $\{\beta_{kl}^h(\cdot)\}$ are convex. Then game (16.9)–(16.12) has a Nash equilibrium contract profile $\mathbf{E}^*$.*

Proof According to [5], an equilibrium exists for any n-person game with concave payoff functions. Since we consider cost functions rather than payoff functions, the same statement is true for games with convex cost functions. Therefore, we need to check whether a cost function $C_i(\mathbf{E}) = C_i(\mathbf{E}_1, \ldots, \mathbf{E}_m)$ is convex in $\mathbf{E}_i$ for each consumer $i \in V_Q$.

Function $C_i(\mathbf{E})$ consists of several summands:

$$C_i(\mathbf{E}) = \sum_{h=1}^{H} \left(G_i^h(\mathbf{E}^h) + T_i^h(\mathbf{E}^h) \right).$$

If we show that each summand in this sum is convex, convexity of the whole sum will be established as well. First, we study function $G_i^h(\mathbf{E}^h)$:

$$G_i^h(\mathbf{E}^h) = \sum_{j=m+1}^{m+n} e_{ij}^h \cdot \alpha_j^h(b_j^h). \tag{16.14}$$

When we fix the contract profiles of all consumers except i, function $\alpha_j^h(b_j^h + \lambda)$ remains convex and increasing, and function in (16.14) is convex as a product of two non-negative increasing convex functions.

Second, we rewrite function $T_i^h(\mathbf{E}^h)$ with fixed contract profiles of all consumers except i:

$$T_i^h(\mathbf{E}^h) = \sum_{(k,l)\in A} \delta_{kl}^{i,h}(\mathbf{E}^h) \cdot \beta_{kl}^h(f_{kl}^h(\mathbf{E}^h)). \tag{16.15}$$

The argument of $\beta_{kl}(\cdot)$ in (16.15) is a linear combination of $\{e_{ij}, j \in V_P\}$, components of consumer i's contract profile. Therefore, $\beta_{kl}^i(\mathbf{E}^h)$ remains convex in $\mathbf{E}_i^h$, as well as $T_i^h(\mathbf{E}^h)$.

The convexity of cost functions in respective arguments is established, that completes the proof.

16.4 Example

Consider a network with seven nodes that is depicted in Fig. 16.1. There are three consumers (red nodes), three producers (green nodes), and one intermediate node. Therefore, $V_Q = \{1, 2, 3\}$, $V_P = \{4, 5, 6\}$, and $V_O = \{7\}$.

Fig. 16.1 7-node network
with no cycles

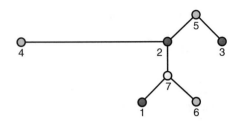

All nodes are located in the same local area except for node 4 that depicts a conventional energy generator, e.g. a power plant. Hence, arc (4, 2) is longer than all other arcs, and transmission costs are higher for this arc.

Since there are no cycles in the network, we only need to check the first Kirchhoff's law. A flow on each arc is a linear combination of $\{e_{ij}^h\}$, $h = \{1, 2, 3, 4\}$:

$$\hat{f}_{25}^h = e_{34}^h + e_{36}^h - e_{15}^h - e_{25}^h, \quad \hat{f}_{53}^h = e_{34}^h + e_{35}^h + e_{36}^h,$$

$$\hat{f}_{42}^h = e_{14}^h + e_{24}^h + e_{34}^h, \quad \hat{f}_{27}^h = e_{14}^h + e_{15}^h - e_{26}^h - e_{36}^h,$$

$$\hat{f}_{71}^h = e_{14}^h + e_{15}^h + e_{16}^h, \quad \hat{f}_{67}^h = e_{16}^h + e_{26}^h + e_{36}^h.$$

The direction of flow in arcs (2, 5) and (2, 7) may differ depending on the values $\{e_{ij}^h\}$. If $\hat{f}_{25}^h < 0$, we assign $\hat{f}_{25}^h = 0$ and $\hat{f}_{52}^h = -\hat{f}_{25}^h$. The same is true for $\hat{f}_{27}^h$.

Let us assume that functions $\{\alpha_j^h(\cdot)\}$ and $\{\beta_{kl}^h(\cdot)\}$ have the following form:

$$\begin{aligned} \alpha_j^h(x) &= \lambda_j^h \cdot x^{1+\epsilon} + \mu_j^h, \quad \forall j \in V_P, \\ \beta_{kl}^h(x) &= \lambda_{kl}^h \cdot x^{1+\zeta}, \quad \forall (k, l) \in A, \end{aligned} \tag{16.16}$$

where all coefficients are non-negative. We are ready now to solve the problem (16.9)–(16.12) with specific values of demands and coefficients in (16.16), and evaluate the total cost reduction. Actually, it is clear that the problem is a computationally difficult. Indeed, the presence of four time periods makes us to compare numerous combinations of different contracts. Thus, we are dealing with combinatorial optimization and the problem could be NP-hard. In future works we will investigate these questions carefully.

16.5   Conclusion

In this work, we have introduced new model for multi-supplier power grid under transmission constraints. Our model studies daily energy dynamics. The game of consumers was formulated and the existence result was established given specific properties of cost functions. There are several directions to improve and generalize the methods discussed in this work, and we name only few of them. First, real-world production and transmission costs, as well as voltage change functions, should be further studied in order to provide realistic representation of the network. Secondly, one can investigate a setting with dynamic network topology. Though power grid structures are relatively constant, there might be different applications of this model, e.g. for planning an optimal modification of a grid, or for maintaining the stability in a case of emergency such as blackouts.

Acknowledgements The author was jointly supported by a grant from the Russian Science Foundation (No. 17-11-01079 | Optimal Behavior in Conflict-Controlled Systems).

References

1. Duffin, R.: Nonlinear networks. IIa. Bull. Am. Math. Soc. **53**, 963–971 (1947)
2. Krylatov, A.Y.: Network flow assignment as a fixed point problem. J. Appl. Ind. Math. **10**(2), 243–256 (2016)
3. Mohsenian-Rad, A.H., Wong, V.W., Jatskevich, J., Schober, R., Leon-Garcia, A.: Autonomous demand-side management based on game-theoretic energy consumption scheduling for the future smart grid. IEEE Trans. Smart Grid **1**(3), 320–331 (2010)
4. Popov, I., Krylatov, A., Zakharov, V., Ivanov, D.: Competitive energy consumption under transmission constraints in a multi-supplier power grid system. Int. J. Syst. Sci. **48**(5), 994–1001 (2017)
5. Rosen, J.B.: Existence and uniqueness of equilibrium points for concave n-person games. Econometrica J. Econ. Soc. **33**(3), 520–534 (1965)
6. Veit, A., Xu, Y., Zheng, R., Chakraborty, N., Sycara, K.P.: Multiagent coordination for energy consumption scheduling in consumer cooperatives. In: Proceedings of the 27th AAAI Conference on Artificial Intelligence, pp. 1362–1368 (2013)

Index

© Springer Nature Switzerland AG 2019
J. B. Song et al. (eds.), *Game Theory for Networking Applications*,
EAI/Springer Innovations in Communication and Computing,
https://doi.org/10.1007/978-3-319-93058-9

Printed in the United States
By Bookmasters